URBAN INNOVATION AND AUTONOMY

Political Implications of Policy Change

Edited by Susan E. Clarke

URBAN INNOVATION Volume 1

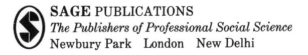

SAGE PUBLICATIONS
The Publishers of Professional Social Science
Newbury Park   London   New Delhi

For information address:

SAGE Publications, Inc.
2111 West Hillcrest Drive
Newbury Park, California 91320

SAGE Publications Ltd.
28 Banner Street
London EC1Y 8QE
England

SAGE Publications India Pvt. Ltd.
M-32 Market
Greater Kailash I
New Delhi 110 048 India

Printed in the United States of America

Library of Congress Cataloging-in-Publication Data

1639654

Main entry under title:

Urban innovation and autonomy : political implications of policy
    change / edited by Susan E. Clarke.
        p.   cm. -- (Urban innovations : v. 1)
    Bibliography: p.
    ISBN 0-8039-3139-5. -- ISBN 0-8039-3140-9 (pbk.)
        1. Urban policy--United States.   2. Urban policy--Europe.
    3. Municipal finance--United States.   4. Municipal finance--Europe.
    I. Clarke, Susan E., 1945-   . II. Series.
    HT123.U7455   1989
    320.8'5--dc20   89-10513
                                                                  CIP

FIRST PRINTING, 1989

Contents

Series Editor's Introduction

The Sage Series in Urban Innovation emerges from the Fiscal Austerity and Urban Innovation (FAUI) Project, which has become the most extensive study of local government in the world. In the United States it includes surveys of local officials in all municipalities over 25,000 population, nearly 1,000. In some 35 other countries, analogous studies are in progress. While project costs exceed $10 million, they have been divided among the participating teams so that some have participated with quite modest investments. Our goal is to document and analyze adoption of innovations by local governments, and thus to sharpen the information base of what works, where, and why. The Project is unusual if not unique in combining a large-scale sophisticated research effort with decentralized data collection, interpretation, and policy analysis. The Project's potential to help cities provide better services at lower costs has heightened interest by public officials. The wide range of survey items makes the data base unique for basic researchers in many related topics. Data are available to interested researchers through the Inter-University Consortium for Social and Political Research, Ann Arbor, Michigan. The Project remains open to persons interested in participating in different ways, from attending conferences to analyzing the data or publishing in the Project's *Newsletter*, its annual volume, *Research in Urban Policy* (JAI Press), or the Sage series in Urban Innovation.

Books in the Sage Series in Urban Innovation may include monographs by a single individual or collective works on a Project theme. The term *fiscal austerity* is not included in the name of the book series because it is not salient to all Project participants; urban innovation is. The availability of a comparable core of data from around the world heightens the international interest even of a volume that focuses on a single country, since teams elsewhere may be encouraged to pursue similar work in other national contexts. Volumes may address a topic in depth in one or more English-speaking countries, or may compare patterns in two or more non-English-speaking countries.

Volumes for the series are reviewed by the Editorial Board: Terry Nichols Clark, University of Chicago, Chair; Harald Baldersheim, University of Bergen; Susan E. Clarke, University of Colorado; Gerd-Michael Hellstern, University of Berlin; David Morgan, University of

Oklahoma; Poul Erik Mouritzen, University of Odense; Robert Stein, Rice University. The Board normally meets with Sage staff once a year in Europe, and once in the United States.

Background

The Project emerged in the summer of 1982. Terry Clark, Richard Bingham, and Brett Hawkins had planned to survey the adaptation of 62 U.S. cities to austerity. We circulated a memo summarizing the survey and welcomed suggestions. The response was overwhelming: People across the United States and in several other countries volunteered to survey leaders in their areas, covering their own costs. Participants were initially attracted by the opportunity to compare cities near them with others. As it seemed clear that we would cover most of the United States, others volunteered to survey remaining states. The result was a network of some 26 U.S. teams using a standard methodology to survey local public officials; the teams pooled their data, and then made the information available to all. The Project spread internationally in the same manner.

While the Project emerged quite spontaneously, it built on experiences joining many participants. Research funds have progressively declined, yet urban research has increased in sophistication and scale. In the past 15 years a few large empirical studies have had major impacts on urban policy analysis. Social scientists and policy analysts increasingly use such studies, but data collection costs are so high that each individual cannot find a grant to collect data he or she might desire. A collective effort thus offers clear payoffs. This situation, recognized in the late 1970s, was the focus of a conference in 1979, where 20 persons presented papers that reviewed the best urban policy research to date, outlined central hypotheses, and itemized critical indicators that might be collected in future work. Seven participants (Terry Clark, Ronald Burt, Lorna Ferguson, John Kasarda, David Knoke, Robert Lineberry, and Elinor Ostrom) then extended the ideas from the separate papers in "Urban Policy Analysis: A New Research Agenda." It was published with the separate papers as *Urban Policy Analysis* (Urban Affairs Annual Reviews, Vol. 21, Sage Publications, 1981). Several persons and many topics from *Urban Policy Analysis* found their way into the present Project.

The Permanent Community Sample (PCS), a national sample of 62 U.S. cities monitored over 20 years, provides a data base and research experience on which the Project built. Many questionnaire items, and methodologies for studying urban processes, were derived from the PCS. Fresh data have regularly been made publicly available; a small data file, provided with a self-instruction manual, has been used for teaching at many universities. Several hundred articles and books have used the PCS; the most comprehensive is T. N. Clark and L. C. Ferguson's *City Money: Political Processes, Fiscal Strain, and Retrenchment* (Columbia University Press, 1983). Basic research and public policy issues have both been addressed, such as how fiscally strained cities are and what solutions they can adopt. These and related issues have been discussed in the United States in conferences, workshops, and publications involving the Department of Housing and Urban Development, U.S. Conference of Mayors, International City Management Association, Municipal Finance Officers Association, and their state and local affiliates. Similar groups have participated internationally, such as the German Association of Cities and many individual local officials. Project participants have come to know each other through professional associations such as the American Political Science Association, International Sociological Association, and European Consortium for Political Research. Meetings in Denver and San Francisco in August 1982 facilitated launching the Project. The international component developed via the Committee on Community Research of the International Sociological Association. This committee had helped organize a conference in Essen, Germany, in October 1981, which led to three volumes published in English by the German HUD: *Applied Urban Research*, edited by G.-M. Hellstern, F. Spreer, and H. Wolman. This Essen meeting and a Mexico City meeting in August 1982 helped extend the Project to Western Europe and other countries.

Since the Project began in 1982, conferences have been held regularly around the world, often with meetings of larger associations, especially the European Consortium for Political Research in the spring and American Political Science Association in the summer.

The Survey:
The Most Extensive Study to Date of Urban Decision Making and Fiscal Policy

The mayor, chair of the city council finance committee, and chief administrative officer or city manager have been surveyed using identical questions in each city of the United States with a population over

25,000 — nearly 1,000 cities. Most U.S. data collection was completed in the winter and spring of 1983. Questionnaires were mailed; telephone follow-ups and interviews were used to increase the response rate. The questionnaire includes items on fiscal management strategies the city has used (from a list of 33, such as contracting out, user fees, privatization, across-the-board cuts, reducing work force through attrition, and deferred maintenance of capital stock), as well as revenue forecasting, integrated financial management systems, performance measures, management rights, and sophistication of economic development analyses. Unlike most studies of local fiscal policy, the Project includes items about local leadership and decision-making patterns, such as preferences of the mayor and council members for more, less, or the same spending in 13 functional areas. Other items include questions on policy preferences, activities, and impact on city government by 20 participants, including city employees, business groups, local media, the elderly, city finance staff, and federal and state agencies. Several items come from past studies of local officials and citizens, thus permitting comparisons of results over time. New data are often shared among Project participants for the first year and then made available to others.

Participants and Coordination

The Project Board, chaired by William Morris, former mayor of Waukegan, Illinois, includes civic leaders and public officials. Terry Clark is coordinator of the Project. Most decisions evolve from collegial discussion. Many participants have 10 to 20 years of experience in working together as former students, collaborators in past studies, and coauthors of many publications. Mark Baldasarre and Lynne Zucker developed the U.S. survey administration procedures. Robert Stein played a leading role in merging U.S. Project data from 26 teams with data from the Census and elsewhere. Paul Eberts is coordinating surveys of counties and smaller municipal governments involving more than a dozen other persons in a closely related study. Participants include persons who helped devise the study, collect or analyze data, or participate in conferences and policy implementation activities. Data collection is complete in the United States and most European countries; it is still under way in some others. Resurveys to assess changes are under consideration.

FAUI Project Participants in the United States

- *Arizona:* Albert K. Karnig
- *California:* Mark Baldasarre, R. Browning, James Danzinger, Roger Kemp, John J. Kirlin, Anthony Pascal, Alan Saltzstein, David Tabb, Herman Turk, Lynne G. Zucker
- *Colorado:* Susan E. Clarke
- *Washington, D.C.:* Jeff Grady, Richard Higgins, Charles H. Levine
- *Florida:* James Ammons, Lynn Appleton, Thomas Lynch, Susan MacManus
- *Georgia:* Frank Thompson
- *Illinois:* James L. Chan, Terry Nichols Clark, Burt Ditkowsky, Warren Jones, Lucinda Kasperson, Rowen Miranda, Bill Morris, Tom Smith, George Tolley, Laura Vertz, Norman Walzer
- *Indiana:* David A. Caputo, David Knoke, Michael LaWell, Elinor Ostrom, Roger B. Parks, Ernest Rueter
- *Kansas:* Robert Lineberry, Paul Schumaker
- *Louisiana:* W. Bartley Hildreth, Robert Whelan
- *Maine:* Lincoln H. Clark, Khi V. Thai
- *Maryland:* John Gist
- *Massachusetts:* Dale Rogers Marshall, Peter H. Rossi, James Vanecko
- *Michigan:* William H. Frey, Bryan Jones, Harold Wolman
- *Minnesota:* Jeffrey Broadbent, Joseph Galaskiewicz
- *Nebraska:* Susan Welch
- *New Hampshire:* Sally Ward
- *New Jersey:* Jack Rabin, Joanna Regulska, Carl Van Horn
- *New York:* Roy Bahl, Paul Eberts, Esther Fuchs, John Logan, Melvin Mister, Robert Shapiro, Joseph Zimmerman
- *North Carolina:* John Kasarda, Peter Marsden
- *Ohio:* Steven Brooks, Jesse Marquette, Penny Marquette, William Pammer
- *Oklahoma:* David R. Morgan
- *Oregon:* Bryan Downes, Kenneth Wong
- *Pennsylvania:* Patrick Larkey, Henry Teune, Wilhelm van Vliet
- *Puerto Rico:* Carlos Muñoz
- *Rhode Island:* Thomas Anton, Michael Rich
- *Tennessee:* Mike Fitzgerald, William Lyons
- *Texas:* Charles Boswell, Richard Cole, Bryan Jones, M. Rosentraub, Robert Stein, Del Tabel
- *Virginia:* Robert DeVoursney, Pat Edwards, Timothy O'Rourke
- *Washington:* Betty Jane Narver
- *Wisconsin:* Lynne-Louise Bernier, Richard Bingham, Brett Hawkins, Robert A. Magill
- *Wyoming:* Cal Clark, Oliver Walter

The non-U.S. participants are among the leading urban analysts in their respective countries, and in several cases direct major monitoring studies with multiyear budgets, including collection of data directly comparable to those in the United States. Gerd-Michael Hellstern, University of Berlin, is coordinating the European teams participating in the Project. Ed Prantilla coordinated the Project in six Asian countries. The survey items are being adapted to different national circumstances, retaining the basic items wherever possible to permit cross-national comparisons.

FAUI Teams Outside the United States

* *Argentina:* Martha Diaz Villegas de Landa
* *Australia:* John Robbins
* *Austria:* H. Bauer
* *Belgium:* Dr. Stassen, Marcel Hotterbeex, Catherine Vigneron
* *Bulgaria:* N. Grigorov, O. Panov
* *Canada:* Andrew S. Harvey, Caroline Andrews, Dan Chekki, Jacques Levilee, James Lightbody, Mary Lynch
* *China:* Min Zhou
* *Denmark:* Carl-Johan Skovsgaard, Finn Bruun, Poul Erik Mouritzen, Kurt Houlberg Nielsen
* *Fiji:* H. M. Gunasekera
* *Finland:* Ari Ylonen, Risto Harisalo
* *France:* Richard Balme, Jean-Yves Nevers, Jeanne Becquart-Leclercq, P. Kukawka, T. Schmitt, Vincent Hoffmann-Martinot
* *Great Britain:* A. Norton, P. M. Jackson, Michael Goldsmith
* *Greece:* Elias Katsoulis, Elisavet Demiri, Clemens Koutsoukis
* *Hong Kong:* P. B. Harris
* *Hungary:* G. Eger, Peteri Gabor
* *Indonesia:* Hatomi, Jonker Tamba
* *Israel:* Daniel Elazar, Avraham Brichta
* *Italy:* Guido Martinotti, Enrico Ercole
* *Japan:* Hachiro Nakamura, Nobusato Kitaoji, Yoshiaki Kobayashi, Yasukuni Iwagami
* *Kenya:* Daniel Bourmaud
* *Netherlands:* A.M.J. Kreukels, Tejo Spit
* *Nigeria:* Dele Olowu, Ladipo Adamolekun
* *Norway:* Harald Baldersheim, Helge O. Larsen, Jonny Holbek, Sissel Hovik, Kari Hesselberg, Nils Aarsæther, Sølbjorg Sørensen, Synnøve Jenssen, Lawrence Rose
* *Philippines:* Ramon C. Bacani, Ed Prantilla
* *Poland:* Gregory Gorzelak, J. Regulski, Z. Dziembowski, Pawel Swianewicz, Andrew Kowalczyk

- *Portugal:* J. P. Martins Barata, Maria Carla Mendes
- *Republic of Korea:* Choong Yong Ahn
- *Senegal:* Abdul Aziz Dia
- *Spain:* Cesar E. Diaz
- *Sweden:* Hokan Magnusson, Lars Strömberg, Cecilia Bökenstrand
- *Switzerland:* A. Rossi, Alberto Naef, Claude Jeanrenaud, Michel Bassand, Erwin Zimmermann
- *Turkey:* U. Ergudor, Ayse Gunes-Ayata
- *West Germany:* B. Hamm, D. H. Mading, Gerd-Michael Hellstern, H. J. Wienen
- *Yugoslavia:* Peter Jambrek

Participation in the Project is relatively open; teams continue to join, especially outside the United States, as they learn of the Project and find ways to merge it with their own activities. Austerity is an issue that links the less affluent countries of the world with others, and one with which the less affluent countries have had more experience. Thus they may be able to offer some distinctive lessons.

Research Foci

Project participants are free to analyze the data as they like, but past work indicates the range of concerns likely to be addressed. The seven-author statement "Urban Policy Analysis: A New Research Agenda" mentioned above outlines several dozen specific hypotheses. Many specific illustrations appear in Project publications, such as the four volumes of *Research in Urban Policy* completed to date; *Urban Innovations as Response to Urban Fiscal Strain* (Verlag Europäische Perspektiven, 1985), edited by Terry Clark, Gerd-Michael Hellstern, and Guido Martinotti; and several country-specific reports. Over 100 papers have been presented at Project conferences, listed in the *Newsletter*. Some general themes follow.

Innovative Strategies Can Be
Isolated and Documented

Showcase cities are valuable in demonstrating that new and creative policies can work. Local officials listen more seriously to other local officials showing them how something works than they do to academicians, consultants, or national government officials. Specific cases are essential for persuasion. But as local officials seldom publicize their innovations, an outside data collection effort can bring significant innovations to more general attention. Relevant questions include the following: What are the strategies that city governments have developed to confront fiscal austerity? How do strategies cluster with one another? Are some more likely to follow others as a function of fiscal austerity? Strategies identified in the survey are being detailed through case studies of individual cities.

Local Governments That Do and Do Not Innovate Can Be Identified: Political Feasibility Can Be Clarified

One can learn from both failure and success. Local officials often suggest that fiscal management strategies like contracting out, volunteers, and privatization are "politically infeasible"; they may work in Phoenix, but not in Stockholm. Yet why not—specifically? Many factors have been hypothesized, and some studied, but much past work is unclear concerning how to make such programs more palatable. The Project is distinctive in probing adoption of innovations, tracing diffusion strategies, and sorting out effects of interrelated variables. Interrelations of strategies with changes in revenues and spending are also being probed.

National Urban Policy Issues

In several countries, and especially the United States, fiscal austerity for cities is compounded by reductions in national government funding for local programs. How are cities of different sorts weathering these developments? Scattered evidence suggests that cities are undergoing some of the most dramatic changes in decades. When city officials come together in their own associations, or testify on problems to the media and their national governments, they can pinpoint city-specific problems. Yet they have difficulty specifying how widely problems and solutions are shared across regions or countries. The Project can contribute to these national urban policy discussions by monitoring local policies and assessing the distinctiveness of national patterns. Fiscal strain indicators of the sort computed for smaller samples of cities are summarized nationally. Types of retrenchment strategies are being assessed. Effects of national program changes are being investigated, such as stimulation-substitution issues. A 269 page report of key national trends in 12 countries has been published by Poul Erik Mouritzen and Kurt Nielsen: *Handbook of Urban Fiscal Data* (University of Odense, Denmark, 1988).

Conclusion

The Project is such a huge undertaking that initial participants doubted its feasibility. It was not planned in advance, but evolved spontaneously as common concerns were recognized. It is a product of distinct austerity in research funding, illustrating concretely that policy analysts can be innovative in the ways they work together. But most of all, it is driven by the dramatic changes in cities around the world, and a concern to understand them so that cities in the future can better adapt to pressures they face. Volumes in the Sage series in Urban Innovation report on these developments.

—Terry Nichols Clark
Series Editor

About the Contributors

Nils Aarsæther received the degree of Dr. Philos. from the University of Tromsø, Norway, in 1986, with a thesis on budgetary behavior under financial strain. Since 1974, he has held Lecturer and Research Fellow positions at the Institute for Social Sciences, University of Tromsø, and in 1984-85 he was Visiting Researcher at the Institute of Political Science, University of Århus, Denmark. From 1976 to 1987, he directed four research projects financed by the Norwegian Council for Research in the Sciences and Humanities. His research has been in the area of local government organization, with an emphasis on development policies, budgeting, and central-local reforms. He has coedited and contributed to several books on municipal and regional development.

Lynn M. Appleton is an Associate Professor in the Department of Sociology and Social Psychology at Florida Atlantic University in Boca Raton, Florida. She has been part of the FAUI Project since the early days of data gathering and cleaning and is glad to have lived long enough to analyze the data. Her current research is on governmental structure and policy outputs, and she is planning work with FAUI cross-national data.

Harald Baldersheim received his Ph.D. from the University of London (LSE) in 1977. He is Professor of Public Administration at the University of Bergen and a Senior Research Fellow at the Norwegian Research Centre in Organization and Management. His recent publications include "When Nobody Wants to Say No" in *Research in Urban Policy* (Vol. 2), edited by T. N. Clark (1986), and several Norwegian-language publications regarding local government in Norway. His research interests include leadership and organizational change in local government, service delivery systems and consumer behavior under nonmarket conditions, and authority structures in private and public organizations.

Richard Balme is Chargé de Recherche for the Fondation Nationale des Sciences Politiques and works with the Centre d'Étude et de Recherche sur la Vie Locale in Bordeaux. From 1986 to 1988, he was a visiting scholar at the University of Chicago and contributed to the comparative analyses of the Fiscal Austerity and Urban Innovation Project. He has published articles on political participation, collective action, and local politics in both American and European journals.

Terry Nichols Clark, as International Coordinator of the Fiscal Austerity and Urban Innovation Project, edits a newsletter and JAI Press annual, *Research in Urban Policy.* He has taught at Columbia, Harvard, Yale, UCLA, and the Sorbonne, and is now Professor of Sociology at the University of Chicago. He has worked at the Brookings Institution, the Urban Institute, the U.S. Conference of Mayors, and the U.S. Department of Housing and Urban Development. He has consulted with many cities on fiscal management questions and has published over 100 books and articles on urban policy issues.

Susan E. Clarke is Associate Professor of Political Science and Associate Director of the Center for Public Policy Research at the University of Colorado. Her articles on state and urban economic development policies, urban social movements, and national urban policy have appeared in American and European journals. Her research interests include urban regime change, political institutions and interest representation, and policies linking economic development and social concerns.

Vincent Hoffmann-Martinot is a Researcher at the Centre National de la Recherche Scientifique in Bordeaux. Along with Jean-Yves Nevers and Jeanne Becquart-Leclercq, he directed the Fiscal Austerity and Urban Innovation Project in France. He is the author of *Finances et Pouvoir Local: L'Experience Allemande,* and his articles on local politics in France and West Germany have been published in American and European journals, including *Les Annales de la Recherche Urbaine* and *The Tocqueville Review.*

Sissel Høvik has a degree in political science from the University of Oslo (1986) and is currently a Researcher at the Norwegian Institute for Urban and Regional Research, working in the field of municipal organization and management. In addition to her contributions to the Norwegian FAUI Project, she is working on projects involving evaluation of a new system of governance in Oslo and evaluation of an experimental program of local government deregulation, with special emphasis placed on the county level.

Anton Kreukels is Professor of Urban and Regional Planning in the Department of Urban Studies at the University of Utrecht. He is also a member of the Netherlands Scientific Council for the Dutch government.

Helge O. Larsen is currently Lecturer (Associate Professor) in the Department of Politics and Administration, Institute of Social Science, University of Tromsø, Norway. His publications include a book on municipal development activity in Norway (with Nils Aarsæther). His current research interests include central-local relations, local government organization and decision-making processes, and, particularly, the study of political leadership at the municipal level. At present, he is working on a project on mayoral leadership.

Poul Erik Mouritzen is Associate Professor in Political Science at the University of Odense, Denmark, and has been a Visiting Scholar at the University of California, Irvine, under the auspices of the American Council of Learned Societies, and at the University of Colorado, Boulder, under the auspices of the Fulbright Foundation. His research interests have included grass-roots political movements, policy analysis, local policymaking during times of fiscal stress, and various aspects of intergovernmental relations. He is currently on leave from the university and working under a three-year research scholarship from the Danish Research Planning Council.

Jean-Yves Nevers is a Researcher at the Centre National de la Recherche Scientifique at the Universite de Toulouse Le Mirail, France. Along with Vincent Hoffmann-Martinot and Jeanne Becquart-Leclercq, he directed the Fiscal Austerity and Urban Innovation Project in France. His articles on local power and politics in France have appeared in American and European journals, including *Les Annales de la Recherche Urbaine, Revue Française de Science Politique,* and *The Tocqueville Review.*

Lawrence Rose received his Ph.D. from Stanford University in 1976. He currently holds positions as Professor of Political Science at the University of Oslo and as Senior Researcher at the Norwegian Institute for Urban and Regional Research. His research interests and publications have covered a variety of topics, including citizen participation, comparative public policy formation, local government, and local community organization. At present he is engaged in two major research projects, both concerning the use of experimental programs as a means of innovation and learning within public organizations.

Tejo Spit is affiliated with the Department of Urban Studies at the University of Utrecht. He is also associated with the Union of Dutch Municipalities.

Ari Ylönen is a Researcher at the Research Institute for Social Sciences at the University of Tampere, Finland, and teaches in the Sociology Department there. He directed the Finnish Fiscal Austerity and Urban Innovation Project, and has published several articles on housing and other urban social issues.

1

Urban Innovation and Autonomy: Cross-National Analyses of Policy Change

SUSAN E. CLARKE

Although a certain urban crisis "nostalgia" (Gottdiener, 1986, p. 273) lingers among urban analysts, it now appears that the scope, magnitude, and persistence of urban fiscal crises in the past decade were not as severe as anticipated by many theorists. But it is also apparent that these phenomena were more pervasive than expected. In varying degrees but with few exceptions, most local officials in the United States and Europe contended with problems of stagnant revenues and constant or increasing expenditure demands over the past ten years. This occurred even in more centralized European countries, where constitutional structures and public finance systems were expected to buffer local communities from the vicissitudes of global economic change (Newton, 1980; Sharp, 1980). Certainly, the magnitude and pervasiveness of local fiscal stress in advanced industrial societies varies considerably across countries, among cities within those countries, and over time. But regardless of the origins or actual degree of local fiscal stress, only a limited number of policy options are possible: increasing resources, limiting expenditures, and enhancing the productivity of the local public sector.

Despite these limitations, a remarkable range of innovative variations on these three policy options is evident in both Europe and the United States. As a consequence, there is growing interest in understanding how, against all odds and theoretical predictions, local officials responded to and seemingly managed the "resource squeeze" (Newton, 1980) of the 1970s and early 1980s. Two key issues are generating important empirical and theoretical work: how to explain

the different types of fiscal policy choices made and how to interpret
the political effects of these choices. Given the initial crisis orientation,
the literature is rife with arguments about the conditions under which
different fiscal policy choices are made, but few of these withstand
empirical analysis (e.g., see Rubin & Rubin, 1986) and even fewer are
borne out in comparative research. Less attention has been given to the
political implications of these policy changes, even though such choices
are at the heart of local allocational processes. There is a sense, as
Gottdiener (1986) puts it, that fiscal crisis alters "the very nature of
local politics" (p. 288), but there are as yet few compelling arguments
about the political effects of local fiscal policy responses.

This volume brings these empirical and interpretive issues together
in presenting initial findings from the Fiscal Austerity and Urban
Innovation (FAUI) Project. In this comparative research project on
local fiscal policy change, researchers in Europe and the United States
used a standardized data collection format and methodological approach
to survey local officials on fiscal change issues. The result is a unique
combination of systematic analysis and country-specific case studies;
it provides reliable, comparative empirical data on local fiscal policy
change as well as the contextual information necessary for interpreting
their theoretical significance. Findings from six countries — the United
States, France, Netherlands, Denmark, Norway, and Finland — are included in this volume. As described below, the Fiscal Austerity and
Urban Innovation Project relies on coordination of numerous autonomous research projects, and these six countries, quite simply, are fur-
thest along in data collection and analysis. The systemic features of
central-local relations in these countries, however, do vary in important
ways; this provides an opportunity to examine intrasystem variations in
fiscal policy choices and to compare these findings across different
systemic conditions.

This first volume in the Sage series in Urban Innovation includes six
comparative case studies analyzing fiscal policy responses within these
countries in terms of a common analytic framework and research
approach. These studies set the stage for subsequent cross-national
comparisons in future volumes by grounding these country-specific
empirical findings in the systemic and contextual features that shape
local policy processes. Data are not pooled across countries or cities,
but the adherence to a common research methodology allows some
comparisons of trends across countries in the concluding chapter. This
introduction presents the rationale for comparative cross-national re-

search on local fiscal policy change and identifies the key issues guiding the initial analyses.

Why a Comparative Approach?

A comparative approach to analysis of these local policy responses is essential for understanding the conditions under which fiscal policy variations occur. It allows us to test various theoretical arguments predicting the nature and direction of policy change and to link local experiences to broader theoretical issues regarding the political effects of policy change. Somewhat paradoxically, the specialized literature on American urban fiscal crisis buttresses the case for a comparative approach. Overall, this literature has evolved in a ratchetlike process: The scope of inquiry has steadily expanded—beyond New York, beyond Cleveland, beyond Rustbelt cities—in order to test specific arguments and increase the generalizability of the findings on fiscal stress. As the scope broadens to include cities sharing similar systemic constraints but with diverse economic, social, and political features, the findings challenge a number of initial assumptions about the nature of fiscal stress and crisis in American cities. The more cities and the more points in time considered in the analysis (e.g., Monkkonen, 1986; Shefter, 1985), for example, the less persuasive the argument that recent fiscal stress in American cities was an aberration that could be laid at the feet of voter-sensitive local politicians (e.g., Auletta, 1979). Nor could it be argued that fiscal stress was an inevitable feature of late capitalism (O'Connor, 1973; Piven & Friedland, 1984) or urban development, since many cities managed to avoid its harshest features (e.g., Merget, 1980). And most striking, there is scant evidence that fiscal stress necessarily leads to fiscal crisis, and growing support for the argument that local fiscal crisis is a political phenomenon and that fiscal stress is amenable to political and managerial approaches (e.g., Rubin & Rubin, 1986; Shefter, 1985; Swanstrom, 1985).

Comparative analyses of local fiscal policy change promise similar salutary results. Such analyses, for example, underscore the anomalous relationship of fiscal stress and fiscal policy change. Haunted perhaps by the specter of New York City's fiscal crisis, early American studies assumed that cities with greater fiscal stress were more likely to seek innovative policy solutions actively. But Clark and Ferguson's (1983) comparative study of 62 American cities disputes the proposition that fiscal policy change is related to fiscal stress: These authors find no statistically significant relationship between the degree of fiscal stress and the adoption of innovative policy changes. Equally telling, similar findings of fiscal policy change independent of the degree of local fiscal stress in European analyses suggest, at the minimum, that fiscal policy change may occur in response to internal interests rather than in response to pressures or shocks external to the policymaking system. This seemingly anomalous relationship between fiscal policy change and local resource conditions in very different local political settings galvanized further comparative research in many FAUI Project sites and raised a number of disquieting issues.

Key Research Issues

The beginning point for each project centers on the barrage of fiscal policy strategies that local administrators and elected officials bring into play when faced with unstable fiscal situations: How do local officials perceive fiscal problems? What choices do they make in responding to these problems? What factors influence these choices? There is a certain fascination in this essentially descriptive exercise. The sheer diversity of strategies tried, the inventiveness and initiative displayed in devising policy responses, the apparent sequencing of strategy choices, and the obvious use of these strategies to reconstruct local coalitions and intergovernmental linkages hints at a degree of local policy discretion, flexibility, and sophistication not often acknowledged by urban theorists.

Policy Innovation and Change Models

These findings on fiscal policy responses brought the assumptions underlying extant models of policy innovation and change into question. The nature of policy change is a central theme in the policy literature. The argument that policy change is associated with a process of analysis, planning, adjustment, and essentially incremental change over time — the technocratic model, as Katznelson (1986) calls it — is challenged by those who see policy change as a response to crisis in the system. The optimistic notion of a smooth, evolutionary change process in a system seeking equilibrium through fine-tuning and adjustments contrasts sharply with the less sanguine view of a system that responds to shocks and disruptions as it must in a process of fits and starts.

These perspectives on policy change reflect political values (Katznelson, 1986), but they also correspond to different views of how organizations innovate and change. As noted above, one set of propositions emphasizes slack-induced innovation, the argument that organizations are most likely and able to change when there are slack resources available to allow experimentation with alternative ways of doing things. The counterargument is that organizations are most likely to change when they have to do so in order to survive or maintain core processes; that is, innovation is induced by stresses and shocks to the organization that compel new processes. In the FAUI Project, the initial assumptions of stress-induced innovation did not withstand systematic comparative analysis, and the potential for slack-induced policy change gained credence. This led to an emphasis on more complex models of fiscal policy change, with greater attention to political leadership and decision-making theories in many analyses.

The multivariate model developed by Clark and Ferguson (1983) served as the prototype, with later refinements (Clark, Burg, & de Landa, 1984) drawing out the effects of local political cultures and political configurations. The effects of special interest groups, especially municipal employees, on fiscal policy choices are of continuing interest, given the importance of personnel costs in local budgets. Yet these multivariate analyses, to date, have yielded relatively modest and unstable findings, especially when used for cross-national analyses (Walzer & Jones, 1989). This suggests further attention to developing the theoretical arguments supporting these models and the consideration of less aggregate measures.

At a minimum, the patterns of diffuse or concentrated political and administrative costs associated with various strategies presumably influence their likelihood of adoption. Some of these patterns stem from the structure of central-local relations: In more centralized systems such as the Scandinavian countries and the Netherlands, local governments might be less willing to cut expenditures in certain areas because clientele and bureaucracies are larger and more entrenched than in countries with less significant welfare systems. For example, Mouritzen (Chapter 4) points out how the welfare state structure in Denmark creates a context in which expenditure cuts are very difficult, since so many local citizens depend directly on state programs or payrolls.

These structural constraints, however, can be mediated by features such as local political party composition. The expectation is that expenditure cuts will vary by party control of local government since parties prefer strategies that mitigate costs for their membership and do not risk incumbency. This argument is one of the most important themes in the public finance policy literature, but it receives mixed support in the studies reported here: It appears to be important in France and Denmark and, to some extent, in Norway, but less important in Finland, the Netherlands, and the United States.

At a different level of analysis, Wolman (1983) posits that strategies are more likely to be adopted if they can be experimented with on a limited basis or are perceived as consistent with existing values and past experiences of local governments. In his comparison of American and British response to fiscal pressures, Wolman depicts local governments as organizations seeking solutions to maintain or reestablish an equilibrium relationship with their environment. When fiscal problems disrupt that environment, he argues that the preferred solutions can be ordered in terms of environmental disruption features. Initial strategy adoptions, for example, would buy time and thus disturb neither external nor internal equilibrium (another interpretation of the managerial response); later, more substantial and disruptive measures include reductions in administrative costs, reductions in real wages, and so on (p. 247).

Local Autonomy

This seeming managerial cast to policy innovation can be interpreted more broadly in terms of the concept of local state autonomy. This presumes that local officials have interests distinct from those of national officials and even independent of dominant local groups. These local state interests include the perpetuation of state institutions and the revenues to support provision of essential services and sustain economic viability as well as short-term concerns with retaining office and maximizing agency resources. In pursuing these interests, local officials can be expected to seek out continually the strategies that allow them to act on their own interests without substantial interference or oversight from these other groups.

Innovative policy choices, therefore, can be interpreted in terms of local officials' institutional incentives to mediate relations among corporate and competitive sectors, central and local government institutions, and economic and social policy concerns in such a way as to increase their relative autonomy from these pressures. Fiscal austerity provides an opportunity to maneuver these relationships and increase local discretionary power and resources. Overall, this argument receives relatively strong support in FAUI analyses to date. There is a surprising similarity in the strategies identified as the most important within each country, independent of the type of central-local relations, political composition of local decision-making bodies, or local pressure group activities. The top-rated strategies tend to be those that allow policy-makers to retain their flexibility and sustain their legitimacy for future decision situations (see Chapter 9).

While local officials' choices can be interpreted in terms of their interests in expanding their ability to take initiatives free from over-sight and interference, the institutional context of these choices is important. Although all local officials may act to enhance their relative autonomy, the degree of local autonomy presumably corresponds to fiscal features of central-local relations. Variations in these systemic features, such as the relative political autonomy and consequent fiscal dependency of American communities, purportedly contribute to fiscal stress and policy responses to that stress. The implication is that com-munities in countries with less decentralized constitutional systems and less restrictive public finance structures — most European countries — would be less vulnerable to fiscal stress than American communities. As these chapters reveal, these presumed relationships among systemic features, fiscal stress, and fiscal policy change are rarely straight-forward.

The concept of relative autonomy is also important in anticipating the political effects of local fiscal policy changes. Do these policy changes lead to an increase or decrease in local communities' relative autonomy from central government controls? And how do they affect local officials' ability to act without significant constraints from local economic and social groups? In the countries studied here, these changes are especially striking. The most important trends include the potential for greater local conflict accompanying decentralization pol-icies, the increased influence of business and taxpayer interests, emer-gent entrepreneurial policy climates, and qualitative changes in local democratic processes. Each of these trends reflects larger changes in these societies and cannot be attributed solely to new fiscal strategies. But local fiscal policy innovations have accelerated or sharpened these trends in many instances. In the concluding chapter, these concepts of autonomy and innovation are used to structure the interpretation of the individual country findings. They prove especially important in think-ing about these political consequences, particularly the potential for democratic local fiscal policymaking in this new economic context.

Note on the Social Production of Knowledge About Local Politics

As everyone knows, the value of attending conferences usually comes from what happens before, after, and between panels. The International Sociological Association meetings in 1981 were no exception to this rule. A number of ISA research committees focus on urban issues, so the numerous panels on urban topics — and the appealing setting — brought a large number of urban researchers to Mexico City in 1981. By organizing the meetings around the traditional siesta, the local hosts created many opportunities for informal conversations among people who, perhaps, knew one another's work but never really had a chance to get together. Stretching over several days of sunshine and sangrita, these discussions inevitably turned to Terry Nichols Clark's novel idea for a research project on local responses to fiscal stress. As Terry described it, this would be an interdisciplinary, collaborative research project in which each researcher would internalize the costs of his or her specific component of the project and all would share the larger data set. All researchers would reach agreement on survey instruments, data collection protocols, data exchanges, and basic research questions to be addressed in the individual projects. Although the initial discussions centered on comparative analysis of American cities, the notion of extending the study to cross-national comparisons quickly took hold. It would be an unfunded, interdisciplinary, cross-national, collaborative, consensual, self-managed, and decentralized research project on fiscal stress.

A more improbable, more infeasible, more unwieldy, more impossible scholarly undertaking is hard to imagine. Everyone said yes, of course, and the Fiscal Austerity and Urban Innovation Project was under way. Over the next few months, the Mexico ambience faded and the illogic of this particular collective endeavor became strikingly clear to some of the original enthusiasts. Surprisingly few dropped out completely, however, and a growing number of researchers who had not attended the ISA meetings expressed an interest in participating. Controlling for sunshine and sangrita, the mutual benefits of participation were clear: In return for conducting a survey on local fiscal stress in their own (and other) states, covering the costs in whatever way they could devise, all researchers would receive comparable data from every other participant. Such a large gain — a standardized data set of survey responses from communities in all 50 states and eventually from Europe, Asia, and Africa — in exchange for a seemingly modest investment of time and resources proved a sufficient incentive for enough researchers to start up the American project. Some people could tap research center or institute funds; others had access to WATS lines, photocopying machines, graduate students, departmental mailing services, and other essential research resources that allowed them to get under way without benefit of external funding. Most of the American studies were conducted by patching together such ephemeral resources; eventually, grants from the U.S. Department of Housing and Urban Development and the Ford Foundation contributed to the costs of coordinating the overall data collection and data management efforts. In both psychic and fiscal terms, however, a good deal of the latter costs were borne by Bob Stein at Rice University, who perhaps unwittingly ended up managing a data archive, clearinghouse, and "control center" for the entire American project.

Several of the European projects benefited from greater external financial and staff support. This is noted by each country team where relevant; such support did not alleviate the frustration and exasperation common to this type of research, but it did allow European researchers to avoid some of the coordination problems hampering the American teams. And in some cases, this support bore a price. The initial Norwegian findings depicted such a positive local fiscal climate that the FAUI Project was blamed by many for subsequent cuts in central government grants to localities. Dutch sponsors, fearing such an outcome, have been especially cautious about allowing release of the survey data, accounting for the somewhat limited findings presented here. Thus even variations in project funding and organization raise issues of autonomy and innovation.

Many, many people ended up spending more time and energy on making the project work than they could ever hope to justify in terms of access to a "standardized, international, urban data set." A putative sense of self-interest certainly was necessary for initial involvement, but it is hardly sufficient to account for the continuing involvement of scores of researchers around the world. Embedded in the FAUI Project are rather inchoate norms of reciprocity and mutual aid that, however imperfectly honored, have resulted in a network of scholars using collaborative, comparative methods to address basic political issues at the local level. As was true of comparative urban research in the 1960s and early 1970s, this new generation of urban research is invigorated by concern with pressing local problems. Years ago, in Mexico City, we started out with the assumption that local fiscal austerity was the issue of the 1980s. But these comparative analyses are reshaping the agenda.

More and more, our concern is with the potential for local democratic
leadership in the face of economic, political, and social pressures
brought about by international economic changes. The possibility of
meaningful local autonomy in the face of these pressures is the issue
for the 1990s.

References

Auletta, K. (1979). *The streets were paved with gold.* New York: Random House.
Clark, T. N., Burg, M. M., & de Landa, M.D.V. (1984). *Urban political cultures and fiscal
    austerity strategies.* Paper presented at the annual meeting of the American Political
    Science Association, Washington, DC.
Clark, T. N., & Ferguson, L. (1983). *City money: Political processes, fiscal strain, and
    retrenchment.* Chicago: University of Chicago Press.
Gottdiener, M. (Ed.). (1986). *Cities in stress: A new look at the urban crisis.* Beverly
    Hills, CA: Sage.
Katznelson, I. (1986). Rethinking the silences of social and economic policy. *Political
    Science Quarterly, 101*(2), 307-325.
Merget, A. (1980). The era of fiscal restraint. In International City Managers Association
    (Ed.), *The municipal yearbook* (pp. 179-192). Washington, DC: International City
    Managers Association.
Monkkonen, E. (1986). The sense of crisis: A historian's point of view. In M. Gottdiener
    (Ed.), *Cities in stress: A new look at the urban crisis* (pp. 20-38). Beverly Hills, CA:
    Sage.
Newton, K. (1980). *Balancing the books.* London: Sage.
O'Connor, J. (1973). *Fiscal crisis of the state.* New York: St. Martin's.
Piven, F. F., & Friedland, R. (1984). Public choice and private power: A theory of fiscal
    crisis. In A. Kirby, P. Knox, & S. Pinch (Eds.), *Public service provision and urban
    development* (pp. 390-420). London: Croom Helm.
Rubin, I., & Rubin, H. J. (1986). Structural theories and urban fiscal stress. In M.
    Gottdiener (Ed.), *Cities in stress: A new look at the urban crisis* (pp. 177-198).
    Beverly Hills, CA: Sage.
Sharp, L. J. (1980). Is there a fiscal crisis in Western European local government?
    International Political Science Review, 1(2), 203-226.
Shefter, M. (1985). *Political crisis/fiscal crisis: The collapse and revival of New York.*
    New York: Basic Books.
Swanstrom, T. (1985). *The crisis of growth politics: Cleveland, Kucinich, and the
    challenge of urban populism.* Philadelphia: Temple University Press.
Walzer, N., & Jones, W. (1989). Fiscal austerity strategies and policy outcomes: A
    comparison of the United States and European countries. In P. E. Mouritzen (Ed.),
    Defending city welfare. London: Sage.
Wolman, H. (1983). Understanding local government responses to fiscal pressure: A
    cross-national analysis. *Journal of Public Policy, 3*(3), 245-264.

2

Coping in American Cities: Fiscal Austerity and Urban Innovations in the 1980s

LYNN M. APPLETON
TERRY NICHOLS CLARK

Urban Fiscal Problems

The New York City fiscal crisis of the early 1970s launched a decade of interest in the fiscal condition of American cities; by the early 1980s, studies of "fiscal stress" and "retrenchment" were staples of both popular and scholarly work. Attention then shifted to coping creatively with fiscal problems. The Fiscal Austerity and Urban Innovation Project began in 1983. In the United States we surveyed mayors, chief executive officers, and council members in all cities over 25,000 population. This chapter outlines the institutional context for studying these issues in American cities, reassesses major theories, and reports some provocative findings from the FAUI Project.

Chief administrative officers (CAOs) were asked the importance of eleven common problems (see Table 2.1). Distant sources were most frequently mentioned: inflation, loss of state and federal revenues, state tax or expenditure limits, and state- and federally mandated costs. Except for municipal employee demands, locally generated problems ranked lower: rising service demands, pressures from local taxpayers, and the like. Of course, defining problems is as much a political process as is their solution, and our respondents' specific positions shape their perceptions. Organizations' environments are "enacted," and "there are as many environments as there are enactors" (Pfeffer & Salancik, 1978, p. 73). Problems identified also reflect the local political culture — the "fact" that employee salaries go up may be labeled "inflation" in Boston but "excessive municipal employee demands" in San Diego. Still, as policy "options" flow from "labeled" problems, we first consider problems and then responses.

Inflation is widely identified as a problem by local officials, even though more careful studies suggest that it often increases revenues as much as expenditures. Net effects are thus usually small (see Clark & Ferguson, 1983; Heidenheimer, Heclo, & Adams, 1983, p. 171; U.S. Congress, Advisory Committee on Intergovernmental Relations, 1979).

Loss of state revenue was a problem for more cities than *loss of federal revenue* — although both are in the top half for our respondents. Why was state more important than federal revenue? State revenues are more often allocated to all cities, so more cities have come to rely on them. Second, state revenues have fewer restrictions, and can thus be used more freely (MacManus, 1978; Stein, 1984; Stein, Sinclair, & Neiman, 1986). Third, states provide cities more money than the federal government, although many "state" funds to cities are "passthrough" funds originating with the federal government.

State limits on taxation and expenditure are often long-standing, but their number and extent grew over the past decade. Some observers (e.g., Pfiffner, 1983) link the "tax revolt" of the 1970s to decline in confidence in government (Janowitz, 1978, pp. 97-113). Others focus on efforts to limit property taxes (e.g., Lucier, 1980), or on "hostility to welfare expenditures from the public purse" (Anton, 1984, p. 31) or reassertion of jurisdiction-based politics (Kirlin, 1984). Still others stress "a middle class populist movement against taxes and spending raised by liberal organized groups" (Clark & Ferguson, 1983, p. 9).

Whatever their origins, these local efforts to "limit government" have spread widely (Pfiffner, 1983). Between 1970 and 1977, 14 states imposed some kind of limit on property taxes; after California's Proposition 13 in 1978, at least 14 others followed suit. By the early 1980s, 29 states had "circuit breaker" protection built into property taxes — rates that drop when the tax is a certain proportion of the taxpayer's income — 9 states had indexed tax brackets to the Consumer Price Index, and at least a dozen states limited state spending.

Municipal employee demands are the main local factor mentioned. Labor costs are the largest item in urban budgets, averaging about half of general expenditures. Early analysts of urban fiscal woes often blamed "organized labor," although more careful analysis limits this impact mainly to compensation (Clark & Ferguson, 1983, pp. 146-165).

TABLE 2.1

City Problems Defined by Chief Administrative Officers

Problem	Percentage of Cities Considering It Serious
Inflation	74.6
Loss of state revenue	50.5
Municipal employee demands	41.0
State tax or expenditure limits	39.1
Loss of federal revenue	36.4
State- and federally mandated costs	34.6
Rising service demands	29.7
Declining tax base	28.0
Unemployment	26.4
Pressures from local taxpayers	26.0
Failure of bond referenda	4.7

SOURCE: U.S. FAUI Project survey, CAO questionnaire.
NOTE: The number or responses varied by item from 484 to 592. Question: "In the last three years, how important have each of the following problems been for your city's finances?" Responses: "One of the most important, very important, somewhat important, least important, don't know/not applicable." The first two responses were considered "serious" in calculating the percentages shown in the table.

State- and federally mandated costs are frequently mentioned (examples include environmental cleanliness standards and buses for the handicapped). Some 75% of state-mandated programs are supported by local revenues (Lovell & Tobin, 1981). While a few states use "fiscal notes" and specific reimbursements for mandated programs, many do not, with the argument that general grants to local governments should cover them.

Rising service demands may flow from population growth (as in Houston), population decline (widely discussed for large, old northeastern cities), or an increase in the proportion of disadvantaged citizens (see Sbragia, 1983a, p. 24; Tabb, 1984).

Most salient for their infrequent mention are *pressures from local taxpayers to cut taxes or expenditures, declining tax base, unemployment*, and *failure of bond referenda*.

Still, we caution that these are reports from administrative staff rather than elected officials, who may define the local policy environment quite differently. (We did not have space in the mayor's questionnaire to ask.) In any case, salience of these problems has a major impact on the strategies adopted, as we will see below.

The Policy Context

What is the "real" policy context for a city? In cross-national discussions, U.S. cities are often said to have great autonomy. But many mayors testify in Washington — when asking for funds — that they are incapable of addressing fundamental local problems. They thus need more federal support. Or when angry constituents or union leaders demand more money, it is convenient for council members to be able to throw up their hands and say, "We can do nothing, it's against the law." Yet these same officials — behind closed doors — can berate each other for not managing local resources carefully, and "not getting the job done" — autonomously. With such strategically misleading information on public record, it is not easy to say precisely what cities can and cannot do. If this problem troubles us in the United States, it seems even worse in Continental European countries, where legally cities can often do less than in the United States, yet a growing body of research suggests that the disparity between those laws and reality may be larger than in the United States.

The standard approaches to autonomy are legal and fiscal. We consider them here, but still stress from the outset that only by analyzing actual behavior in decisions can one assess autonomy as it affects concrete decisions. Consider a strong test case, strong in that it involves laws and money (along with other variables): debt and tax limits. The amount of debt a city may issue and the tax rate are limited by law in most states. But studies of actual debt and taxes have increasingly concluded that the legal limits exercise little or no effect on the actual amounts of debt issued (Hickham, Berne, & Stiefel, 1981). Smart local officials can work around many laws, or help pass new ones, or find administrative solutions to do what their constituents want. Similarly, they can pursue federal and state grants for purposes their constituents support, but not others. Thus both laws and intergovernmental grants may be "adapted" — to some degree. Precisely how much is a topic over which legitimate disagreement persists. We review some evidence below, but stress simply that the more one moves away from legal and fiscal definitions and analyses, and the more one studies specific decisions in their local context, the more cities seem autonomous. A similar recognition of greater local autonomy has occurred in Europe over the past 20 years as studies have become more informed by actual local dynamics (see Clark, Burg, & de Landa, 1984; Thoenig, 1986).

Autonomy appears, or disappears, as cities meet constraints. We consider first constraints from state and federal governments, and, second, local political and economic constraints on policy responses by local governments.

Extralocal Constraints

Federal constitutional authority is divided between the federal and state governments; cities have no constitutional role. Thus cities are "creatures of the state" and (50 different) state laws and regulations define most specifics for local governments, making for considerable diversity across the country. Preeminence of the state, and legal silence about the federal government, characterized intergovernmental relations from the eighteenth until the early twentieth century. Until the 1930s, cities received little money from either federal or state governments. Inequalities in cities' abilities to provide services were considered an inevitable and perhaps desirable concomitant of a "free market" economy and a federal system. Still, the U.S. Constitution stresses both equal rights of citizens and liberty to pursue their individual goals; the tension between these two basic values has generated much comment. By contrast, some Continental constitutions, such as France's, stress equality more. Certainly, the American federal system did not encourage equalization of resources across jurisdictions (Elazar, 1972; Hamilton, 1980). The political parties have similarly reflected local diversity.

Franklin Delano Roosevelt changed things in 1933 by mobilizing many unemployed, poor, and immigrant citizens — traditionally low or nonvoters — bringing them into political awareness (and into the Democratic party) by offering a "New Deal." As Benton and Morgan (1986) put it, "In only a few short years, governments were transformed from entities that best fit the exhortation of Jefferson (i.e., government which governs best, governs least) to what many have variously called the positive or welfare state" (p. 7).

Yet which programs should grow, and how, was a source of continual political battle. Mollenkopf (1983) reviews these battles in left-right terms, suggesting that sometimes Congress and the president leaned left, expanding redistributive aid to the urban poor. But this in turn brought counterreactions by the right. Between 1965 and 1968, over 200 new federal programs were launched under Democratic President

Lyndon Johnson. The most dramatic (such as drug rehabilitation centers, or War on Poverty requirements of maximum feasible participation by program recipients) fueled political response. Republican Presidents Richard Nixon and Gerald Ford (1968-1974) sought to dismantle such programs. Yet because Congress remained strongly Democratic, compromise was necessary. The most common tactic was to consolidate a dozen or so controversial individual programs into general "block grants," such as the Community Development Block Grant of 1974, and permit more local autonomy in their administration. This compromise meant that more poor-oriented cities (like New York) could continue programs with federal funds, while others could shift funds to other purposes if they so chose. Compromise generally meant more funds for each interested party, however, and in the 1970s the term *special interest group* entered political discourse to refer to those groups that bargained additions into federal programs to their own advantage. The rise of such groups was also linked with weakening of the national parties. Always weak by European standards, 1972 party reforms weakened them further: Seniority was abolished as the key criterion for selecting leaders in the U.S. Congress. Many young members of Congress then sought some basis to distinguish themselves. They often allied themselves with leaders of distinct organized groups in Washington. Party leaders thereafter could not control the dozens of young members of Congress and organized groups as in earlier years. Blacks, feminists, ecologists, unionists, the elderly — all went their separate ways. Despite Republicans in the White House in the early 1970s, and then the fiscally conservative Democrat Jimmy Carter (1976-80), direct federal aid to cities grew 268% from 1972 and 1977, and federal aid to the states that "passed through" to cities grew 73%.

These are aggregate results, however — national averages. They conceal huge differences across cities. About 80% of federal grants remain contingent on specific applications from the local government, and are awarded for specific purposes and projects. In the 1960s and 1970s, cities like New Haven, Boston, and New York developed programs locally that they then took to Washington and often found congressional sponsors that converted them into federal programs. Brilliant success at such "grantsmanship" was illustrated by legendary figures like Edward Logue, director of urban renewal in New Haven (see Dahl, 1961) and then Boston. Mayors and councils, of course, appointed grantsmen in such cities. The importance of this kind of political support is clear when one contrasts these cities with Houston, San Diego, and many smaller municipalities that were much more cautious about seeking federal grants. When federal and state grants were being cut in the late 1970s, the San Diego City Council commissioned a report on San Diego's dependence on higher-level governments:

> We concluded that the City really is not very dependent upon State and Federal Government funds and that loss of these funds would not seriously affect City services. When one reads the "horror" stories of other cities, it is comforting to find that San Diego has refused to accept the carrot of easy Federal and State funds. (San Diego Board of Fiscal Overseers, quoted in Clark & Ferguson, 1983, pp. 66-67)

These differences in orientation led to large differences in allocation of, and dependence on, intergovernmental funds. How much such "dependence" was in fact a local choice grew clearer as some federal programs were cut in the late 1970s. Social service programs were often deemed essential in cities like New York, which continued them with local taxes. By contrast, Chicago as a general policy terminated such programs. But cities like San Diego or Houston, which had never applied to participate in many programs in the first place, had nothing to cut. Differences across these three kinds of cities illustrate the more general conclusion that emerges from several studies of the determinants of federal grant allocations, and their effects on local spending. From the 1960s until the mid-1970s, federal and state grants to cities were awarded so much due to grantsmanship that various poverty and "need" measures of cities were almost uncorrelated with intergovernmental grant receipts. This changed somewhat in the later 1970s as allocation formulas came to be more widely used. Similarly, intergovernmental grants (as independent variables) minimally explain differences in local spending patterns across cities; most spending differences are due to local citizen and leader characteristics. Cities with fiscally liberal citizens and leaders spend more of their local tax money; they also encourage staff to write many applications and obtain intergovernmental grants (Clark & Ferguson, 1983, chap. 9).

How does this picture of local autonomy appear when, instead of spending growth, we shift to drastic spending cuts? Just when and where cuts occurred are important to examine for an answer. Many Americans date the taxpayer revolt by the 1980 election of Ronald Reagan as president. Others see it in the 1978 passage of Proposition 13 in California, which cut property taxes and limited spending. But as Clark and Ferguson (1983, pp. 49) show, the revolt began four full years before Proposition 13, and at the local level, when about half of American cities began trimming their own budgets. These were local political decisions, implemented largely by new mayors and council members who were fiscally conservative Democrats. And they were elected, and voted such budget cuts, because their citizens spoke out for them, in hearings and surveys. National policy thus chronologically followed local policy in this most important fiscal policy shift since the 1930s. The taxpayers' revolt was not imposed on cities; cities started it. This again is a critical demonstration of local autonomy.

Seen in international context, the famous tax revolt also emerged from classic American beliefs about the virtues of individualism, local community self-sufficiency, and distrust of bigness. The New Deal and its Great Society legacy remained politically controversial into the 1970s. Some city councils adopted laws prohibiting their staffs from seeking or obtaining any grants from the federal government — they fought "creeping socialism." Many cities accepted their first federal grants only in 1972, when the first General Revenue Sharing checks were sent out. American local officials have far more horizontal contacts than French officials, who have more vertical contacts with the national government (Balme, Becquart-Leclercq, Clark, Hoffmann-Martinot, & Nevers, 1987). Such American localism also means that local officials pose a less potent electoral threat to national leaders than in more nationally integrated systems. Since few American local officials can "deliver the vote" in national or local elections, they cannot go to Washington to claim the rewards of doing so. (There are famous exceptions, of course, such as Chicago's Mayor Daley.)

While citizens used special interest groups to lobby in Washington, local governments used "public interest groups" (PIGs). The most important of these was the U.S. Conference of Mayors, a core actor in expanding the New Deal legacy through the 1970s. Republican mayors seldom participated in it until after the 1980 Reagan election; it then declined as a lobbying group. Reagan appointments to agencies like the Advisory Commission on Intergovernmental Relations also had an impact. These changes marked a definite shift in tone of much public discussion of national urban issues.

In the late 1980s, with a Republican White House and states still adjusting to the tax revolt, cities lobby less and less successfully for intergovernmental support. The most popular programs are those accessible to the majority of American cities, those without substantial dependent populations or decaying infrastructures. Such programs have been damaged least by the 1980s cutbacks in federal aid. The most visible and pointed cuts have been in social services. Drastic cuts were proposed in the early Reagan years, including a complete reallocation of many federal functions back to state and local governments. But this "New Federalism" was rejected by Congress, which also restored many cuts proposed in the president's budget. Precisely how large the cuts are, who is responsible for them, and how much they are supported by state and local officials as well as citizens remain controversial. When the feds cut, states grudgingly "picked up the tab" for selected programs (Nathan & Doolittle, 1983), but the states by no means made up the gap left by the federal retreat.

While there is still a welfare state firmly in place, it is by no means unshaken. Many individual programs, especially those most clearly oriented toward the disadvantaged, have been abolished or consolidated into broader block grants in the early 1980s. And the total volume of funding is clearly down. Most painful to mayors, however, was termination of the General Revenue Sharing program, which for some 15 years had provided about 10% of total federal revenues to municipalities. Mayors particularly liked Revenue Sharing, as it was the only program that could be spent almost entirely at local discretion. It seems to have been abolished in part since mayors and other local officials were less successful than other groups in developing a strong constituency with Congress.

While many inflationary, trade, and other economic problems are shared by other countries, intergovernmental grants were cut more in the United States than in other countries in Europe or Japan (Mouritzen & Nielsen, 1988, p. 38). Intergovernmental grants from federal and state governments to local governments peaked about 1977, and have declined since. As a percentage of all local revenues, intergovernmental grants rose from 24% in 1970 to 32% in 1975, and then fell back to 24% by 1985 (Table 2.2). State grants appear almost twice as large as federal grants, but a good portion of state funds for cities originates with the federal government. There are many ways to compute such figures, of course. For example, if one omits utilities and insurance trust (pension) revenues from the Table 2.2 totals, and then calculates intergovernmental grants as a percentage of own revenues, the drop from 1977 to 1986 is almost half.

While the growth and decline of federal aid to cities has been much discussed, state actions are often more important for cities — as was clear in Table 2.1. And the states provide almost twice as much money as the feds (ignoring pass-throughs; Table 2.2). A related development in the past few decades has been minimally recognized: States have been granting local governments more autonomy. The most visible legal package is *home rule*, which implies legal local autonomy in all

TABLE 2.2

Average Local Revenues, by Source

	1969-70	1974-75	1979-80	1984-85
General revenue				
intergovernmental				
from state government (%)	18.9	21.8	16.8	15.6
from federal government (%)	4.1	9.8	11.5	7.0
from local government (%)	1.2	1.3	1.5	1.7
subtotal (%)	24.2	32.9	29.8	24.3
own sources				
taxes (property, sales) (%)	41.7	35.4	32.9	32.3
charges and miscellaneous (%)	15.5	15.2	17.4	21.1
subtotal (%)	57.2	50.6	50.4	53.4
Utility revenue (%)	15.4	13.8	16.3	17.8
Liquor store revenue (%)	0.4	0.3	0.3	0.2
Insurance trust revenue (%)	2.8	2.5	3.3	4.4
Total (%)	100.0	100.0	100.0	100.0
Total revenue (millions of dollars)	32,704	59,745	94,863	147,648
Average total revenue (dollars per capita)	201.82	320.62	539.56	781.34

SOURCE: Computed from U.S. Department of Commerce (1986, Tables 1,4).
NOTE: Data for all U.S. city governments. See note 1.

functional areas. Yet what home rule specifically means varies considerably across states and cities, and over time. With or without using the legal term *home rule*, many states delegate considerable autonomy to their local governments. Nevertheless, two areas where state law directly constrains local options are in *revenue structure* and *local functional responsibilities*.

The largest local revenue is the property tax (see Table 2.2 and Figure 2.1).[1] States vary widely in their restrictions on property taxes and willingness to grant exceptions (MacManus, 1983, pp. 151-154). They may limit the tax rate, its rate of increase, publicity surrounding its increase, the rate of increase in property assessments, and types of property tax exemptions. These constraints have encouraged cities to seek other revenue sources. Nonproperty taxes are formally more constrained; most states restrict taxes on income, sales, or privileges. The least constrained local revenues are charges (e.g., user fees) and miscellaneous local revenues (e.g., special assessments, sale of city prop-

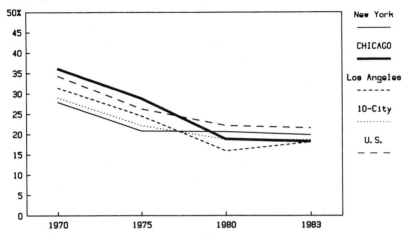

Figure 2.1 Property Taxes as a Percentage of General Revenues, 1970-1983

SOURCE: Clark (1985, p. 26), compiled in turn from U.S. Bureau of the Census, *City Government Finances,* annual series.

erty, interest earnings), hence the increasing popularity of these mechanisms. These last revenue sources have grown increasingly popular over the last decade (Rigos, 1986).

Local Constraints

What local constraints shape policy options for American city governments? The economic base, citizens, organized groups, and city staff are seen as particularly important features of American local politics.

Economic constraints are often cited first. While debate on economic versus political explanations of public policy has left a significant literature, it is clouded by shifting definitions of what each is, and how specific processes operate. An economic determinism of local politics has characterized much work, from Lynd and Lynd's classic *Middletown In Transition* (1937) through Floyd Hunter's *Community Power Structure* (1953). In comparative work in the mid-1960s, this led Thomas Dye, Richard Hofferbert, and others to contrast socioeconomic variables such as population size, wealth, and race with political variables such as form of government and political party in regressions to explain public expenditures. They reported more impact by socioeconomic variables.

Another economic constraint is competition by cities for firms and residents. Peterson (1981) argues that such competition essentially prohibits local governments from pursuing redistributive policies. Stone (1980) similarly argues that economic interests shape city policy, even when economic actors are not overtly involved in politics. To the degree that city governments depend upon their local resource base to fund their activities, they should thus adopt policies that enhance — or, minimally, do not change — that base.

Research on urban policy outputs, however, has shown economic factors less powerful: Politics matters. The picture shifted first as more refined indicators and analyses (e.g., interactions as well as direct effects) were used. Second, sharper conceptual work showed the political meaning of variables like income and race — they measure both resources (a traditional economic variable) and policy preferences (a more political variable). Proper model specification can discern each effect. Such analyses led Clark and Ferguson (1983, chap. 3) to critique the "Northeast syndrome" explanation of fiscal strain — which saw a declining economic base as straining local government. They show that fiscal strain is only weakly related to economic decline and similar population characteristics.

How about *organized groups*? From Arthur Bentley's work in the 1920s through David Truman's *The Governmental Process* (1951), there emerged the mainstream political tradition of "group theory." It held that in the beginning there were interests, and that organized groups were the primary actors pursuing these interests in politics. Robert Dahl elaborated this approach into his "elitist theory of democracy," which similarly assigned group leaders and elected officials critical policy roles (see, e.g., Dahl, 1961). In cities and time periods (about 1970) where organized groups were active, they also increased city government spending "out of proportion" to their local resource base. Still, interest organization varies across cities and issues. While Yates (1977) characterizes American city politics as "street-fighting pluralism," Peterson (1981, pp. 116-128) dwells on the absence of ongoing organized interest groups or strong political parties in most American cities. If the 1960s saw intense group involvement in some cities, organized activism is exceptional. Local electoral participation is low (Verba & Nie, 1972), especially among the poor and near poor. And voters have grown increasingly alienated from and distrustful of political institutions (Janowitz, 1978, pp. 97-113). In most cities, the local political environment is neither highly organized nor polarized. The majority of citizens are not involved or interested in city government decision making; they mobilize rarely. Most decisions of city government, then, are based on little direct information about "what the people want" (Clark, 1976). Elected officials after the late 1970s have felt that most of their constituents dislike tax increases — but, beyond that, they often know little.

While *citizens* have always been the source of authority, they were long seen as "expressing their preferences" indirectly. This changed in the 1960s and 1970s, when citizens became a far more visible and direct force in American cities — first with citizen participation in the War on Poverty programs, and then in the taxpayer revolts. These changes have led researchers to look more closely at citizens' preferences and modes of policy input. Citizens constrain policymaking, but researchers (and policymakers) need sensitive measures to assess such constraints. For example, over the 1960s and 1970s, spending increases were more likely in cities with smaller middle classes, more black residents, and income-heterogeneous populations. Surveys of citizens similarly show that middle-class Americans prefer less spending, while poor and black Americans prefer more.

How about policy constraints from *city government staff*? Because of weak political parties and (usually) silent citizens, city officials often lack clear methods to define popular preferences. Some hold that even if elected officials were clear about those preferences, not all would feel constrained to act on them. American politicians are often part-timers without political career ambitions beyond the city, and remain enmeshed in private concerns. Further, the structure of city government often encourages elected officials to avoid specific policy decisions. Between one-third and one-half of U.S. cities have elements of "reform government," such as at-large and nonpartisan elections. In such cities, a city manager is normally responsible for operating decisions, while the mayor and council are part-time officials with limited authority. Legally "weak mayors" do not appoint department heads (the manager does), and staff report to the council via the manager. It is often illegal for the mayor even to phone a city staff member for information; he or she must pass through the manager. This conscious effort to isolate "government" from "politics" comes from the late nineteenth-century movement of (often nativist Protestant) Americans fighting the patronage or clientelism of "corruption" often identified with immigrant Catholics. Even cities with formally strong mayors have developed large bureaucracies that in turn often weaken political control. Elected officials tend to defer to professional expertise. Still, the mayor and council retain a critical resource: They hire and fire the manager. The "amateur politician" image is an American tradition contrasting with that of the European professional (although Larsen, 1987, stresses the amateur leadership of many smaller Scandinavian towns). This amateur tradition seemed stronger before the tumultuous 1960s, when many amateurs returned to their country clubs, leaving politics to the more serious (contrast the quiescent image of California cities in Eulau & Prewitt, 1973, with the turbulent portrayal of these cities in Browning, Marshall, & Tabb, 1984). Still, specific policy roles of elected and appointed officials remain unclear in much past literature. This ambiguity led us to include several items in the survey.

Fiscal Strategies

Table 2.3 lists the 33 strategies from our survey. There are many ways to classify the 33. First are strategies that eliminate staff, or second visibly change the service mix. Both would arouse opposition. Third are those that staff may prefer to increase organizational domains or control over subordinates, even if this hurts service recipients. Fourth are strategies that avoid legal constraints: States have few rules about own-source-generated user fees (ranked first) and other nontax revenues (e.g., "new local revenue sources," ranked third). Fifth are strategies that affect personnel, which can be adapted to civil service and union constraints; for example, attrition (ranked fourth) is usually easier to implement than early retirement (ranked thirty-first).

Consider user fees. They stand first in Table 2.3, for reasons that illuminate American policy preferences. Fees are widely considered "fairer" than taxes since individual users pay for services received — marginal supply is thus matched with marginal demand and "economic efficiency" realized, the clinching argument in neoclassical economics. Fees also provide information about citizen preferences (Mushkin & Vehorn, 1980). These aspects of fees appeal to American individualism, which in the last decade has swamped the egalitarian argument that all citizens should receive the same services. Some evidence suggests that fees are adopted by fiscally healthy cities with a significant middle class, which prefers them to taxes (Rigos, 1986). Yet there are important exceptions, such as Democratic strongholds like Chicago, where egalitarian arguments may win over the individualistic.

Other reasons for fees are more general. They can help the department offering the service: If citizens pay for garbage pickup, the sanitation department receives the "benefits." Many states restrict fees

TABLE 2.3

Percentage of U.S. Cities Using Each of 33 Strategies (N = 517)

Strategy	Percentage of Cities Using Strategy
Increase user fees	85.1
Improve productivity through management techniques	74.6
Add new local revenue sources	73.3
Reduce work force through attrition	70.9
Reduce spending on supplies, etc.	67.7
Draw down surpluses	64.2
Improve productivity through labor-saving techniques	63.0
Increase taxes	60.5
Hiring freeze	58.9
Reduce capital expenditures	58.4
Across-the-board budget cuts	56.0
New intergovernmental revenue	53.9
Reduce overtime	53.3
Reduce administrative expenses	52.4
Keep expenditure increase below rate of inflation	50.6
Eliminate programs	47.5
Contract services out to private business	45.8
Joint purchasing agreements	45.6
Layoffs	45.6
Sell assets	41.2
Reduce locally funded services	39.6
Cut least efficient departments	39.4
Defer maintenance	36.9
Reduce intergovernmental-funded services	36.5
Increase long-term borrowing	35.7
Defer payments	35.0
Increase short-term borrowing	34.2
Shift responsibilities to other governments	31.7
Freeze salaries	31.1
Contract services out to other goverments	28.6
Early retirements	26.1
Reduce pay levels	23.0
Control new construction to limit population	18.9

SOURCE: U.S. FAUI Project survey, CAO questionaire.
NOTE: Item: "Here is a list of fiscal management strategies that cities have used. Which have you used since January 1978? . . . Please indicate the importance in dollars of each strategy (since January 1978): one of most important, very important, somewhat important, least important don't know/not applicable."

to the costs of service (MacManus, 1983, p. 160). Yet, since "actual cost" involves some discretion, department heads may use fees to display their revenue-raising talents, expand their organizational domains, and advance their careers.

Cities seldom adopt innovations that concentrate costs on a group capable of concerted political action, hence a reluctance to trim functional responsibilities by eliminating programs or to shift programs to other governments. This also avoids conflict with city government staff having a stake in the program. The importance of not antagonizing staff is underscored by the unpopularity of strategies materially affecting municipal employees: reducing pay, early retirements, freezing wages, layoffs. When spending cuts are adopted, they are far more often across-the-board than concentrated in the least efficient departments. This disperses, rather than concentrates, resistance. Ceteris paribus, across-the-board strategies should be more important in cities (and countries) where municipal employees are more powerful, while more targeted cuts should occur where those employees are weaker.

Finally, few cities increase long- or short-term borrowing or defer payments on existing loans, as New York did in its fiscal crisis. This reflects post-Proposition 13 state legislation limiting such practices (MacManus, 1983), local conservatism, and reluctance to appear fiscally irresponsible to investors, which would increase interest costs (Sbragia, 1983a).

Explaining Adoption of Fiscal Strategies

We have so far stressed rationales similar for most cities. But why do cities differ in use of strategies? The 33 strategies can be divided into three types: (1) those that increase revenues, (2) those that decrease expenditures, and (3) those that increase productivity. Our three dependent variables (in Table 2.5, discussed below) simply sum the number of strategies of each of these three types used by the city.[2]

Can we rank the three types on innovativeness — at least as departure from custom (Mohr, 1982)? Considered as breaking from practice in recent decades, increasing revenues would be least innovative, productivity medium, and cutting expenditures most distinctive. Still, productivity may be the hardest to implement due to organizational resistance by city staff, even if it responds most to citizen and taxpayer demand.

Sources of the three types of strategies were first reported in Clark et al. (1984). Six subsequent studies of the same strategies were completed by other participants in the FAUI Project, summarized in Table 2.4. Further ideas come from analogous studies in countries outside the United States, represented by chapters in this volume. To facilitate cross-national comparison we have replicated the Norwegian regression analysis and part of the French analysis.[3]

What emerges? We consider results by type of explanatory variable, drawing on Table 2.4 for the earlier studies and Table 2.5 for our current regression results.

Should *fiscal strain* affect choice of strategies? Contrasting hypotheses are the "incrementalist" view that decisions change slowly (even fiscally strained governments will continue past policies; Wildavsky, 1964); the simple "rationalist" argument that fiscally strained cities will be forced to adopt more fiscal austerity measures than less strained cities (Scott, 1981, especially chap. 2); and the "organizational politics" argument that only fiscally healthy cities can innovate since innovation requires organizational slack (Levine, Rubin, & Wolohojian, 1981). Each proposes a relationship between fiscal stress and policy adoption.

Instead, we hypothesize no relationship. City government is not a mechanism, like a closed black box, where stress leads directly to specific policies. Fiscal stress is bound up with other changes: Government structure and culture affect when and which fiscal problems are ignored or stressed. Even cities with histories of reckless fiscal practices have reversed these, especially when candidates for mayor and council have challenged the status quo, won, and then changed policy dramatically (e.g., New York, Pittsburgh, and many smaller cities in the 1970s).

We define fiscal strain as lack of adaptation by government to a changing environment, operationalized by a ratio of expenditures on nine functions common to most American cities in 1980, over per capita income in 1979 (U.S. Department of Commerce, 1983). For this and most population measures, 1979 is the latest available year, but it is also appropriate since it is near the start of the period about which we asked in the survey.

We find, as expected, no relationship between fiscal strain and revenue, expenditure, or productivity strategies. We similarly found no relations between the strategies and alternative fiscal measures: change in own revenues, general expenditures, debt or common functions from 1980 to 1984. (Appleton, 1988, considers nine such measures.)

TABLE 2.4

Sources of Productivity Strategies in Seven Past U.S. Studies
from the Fiscal Austerity and Urban Innovation Project

	CBL	CW1	CW2	B	MP	A	H
R		.36	.49				.50
R^2	.14			.13	.13	NA	.25
Adjusted R^2					.09		
Independent Variables							
Fiscal strain							
expend/PC income	0	0	0	0			
change in revenue 75-82					.14	low	
change in debt 75-80					-.12	low	
Political culture							
Democratic	0				.09	varies	
Republican	0				0	by	
Ethnic	-.20		0	0		period	
New Fiscal Populist	.16		.14	.13		+ and -	
Mayor							
Democratic party	-.17			-.10			
fiscal liberalism		0			0		
Fisc Lib*Mys Power	.16		0				
Organized groups							
active liberal groups			.16		0		
active Democratic party			-.12				
Salient problems							
all					0		
federal cuts						.14	.17
state cuts		0	0			.18	-.26
inflation, etc.						.13	
decline/unemployment		0	0			.17	-.03
municipal employees		.15				.09	.39
taxpayers			.26			.18	.39
Socioeconomic characteristics							
city size		0	0				
population change			0				
per capita income		.02	0	0			-.03
foreign stock	0				0		
percentage black	0				0		
percentage minority			0				
percentage poor	0						
manufacturing, age, and region					.23		

TABLE 2.4
(continued)

	CBL	CW1	CW2	B	MP	A	H
Administrative staff							
city manager government	0	0	0	0	0		
fiscal management sophistication		.27	.25			.21	high
staff power	.26		0	.29			
Other							
January temperature			0				

NOTE: Most studies classified the 33 strategies into the same three types (revenue, expenditure, productivity). Most used the same independent variables for all three types of strategies, although this table reports results only for the number of productivity strategies adopted by the city. All studies (except A) use OLS regression. The R, R^2, and adjusted R^2 at the top are for the full equation as reported in each study. Coefficients below them are betas, reported only if over .10 significance, otherwise shown as 0 if included in the model, or left blank if not included. For example, the CBL study used 12 independent variables generating an R^2 of .14; the beta for its fiscal strain measure was 0. Some studies combine items into indexes; the main item is then listed. A and H are distinctive enough in method from others that readers should consult the original papers. Studies: CBL = Clark, Burg and de Landa (1984); CW1 = Clark and Walter (1986) (dependent variable is "professional retrenchment" index, summing 15 strategies); CW2 = Clark and Walter (1987); MP = Morgan and Pammer (1986) (see also Pammer, 1986, for more detail.); B = Burg (1986) (a refinement of CBL); A = Appleton (1988) (reports r's with various controls included; distinguishes strategies by year; NFP political culture increases productivity in early years, opposite later.); H = Hawkins (1988). Most variables from Hawkins are indexes of 4-7 components, sometimes reported twice in this table if two components in one index, e.g., municipal employee and taxpayer problems. Higher R^2 may be due to lower N.

A related measure is *percentage of intergovernmental revenues*, which was generally unrelated to strategy adoption. It was thus omitted from Table 2.5, but does appear in our Norwegian replication in the Appendix to this chapter. Clark and Walter (in press) explored intergovernmental revenues in depth and found several interesting relations between intergovernmental revenue patterns and FAUI survey items.

The *tax base* we measure using per capita income (1979) (U.S. Department of Commerce, 1983), the most comprehensive measure and frequently used in research. It lies behind many of the sources of local revenue in Table 2.2. If fiscal problems are largely tax base problems — as the economic determinist argument of Dye and others suggests — then the tax base should largely explain strategy adoption. But it does not at all: Per capita income is simply unrelated to the strategies.

Government structure, as discussed above, has been studied frequently. We include the key "reform" measure of having a professional

TABLE 2.5
Sources of Strategies: Reduced Form Regression Model

	Revenue Raising	Dependent Variables: Strategies for Expenditure Reduction	Productivity Improvement
Multiple R	.350	.424	.455
R²	.122	.179	.207
Adjusted R²	.088	.147	.176
Independent Variables			
All problems	.178**	.247**	.121*
Mayor spending*my influence	.149*	.148*	.158**
Fiscal management index	.023	.066	.311**
Functional performance index	.189*	.158*	.178**
Population black 1980	.065	.068	.035
Proportion Democratic council members	−.181**	−.207**	−.224**
Population 1980	.068	.102	−.015
Common funcs80/per capita income	−8.4E−03	.059	.032

NOTE: This table presents standardized regression coefficients (betas) estimated using OLS procedures. This reduced-form model is simplified from the structural model in note 3 by deleting insignificant variables. All variables are logs and are discussed in the text. Data from U.S. FAUI Project survey, including survey data for cities that were not available before fall 1987 from 11 states. Pairwise deletion of missing cases is used to preserve as many cases as are available for each bivariate correlation from which the regression is calculated. The lowest bivariate N is 213.
*Significant at .05 level; **significant at .01 level.

city manager (from the International City Management Association's *Municipal Yearbook*). While some early comparative studies found that reformed governments differed in their policies (Lineberry & Fowler, 1967), later studies report mixed findings (Abney & Lauth, 1986; Morgan & Pelissero, 1980). Still, a working hypothesis is that reformed city governments will adopt more strategies than unreformed cities because more apolitical administrators are more responsive to professional appeals for innovation. While this is plausible, it is not supported. All five past studies including reform government (in Table 2.4) found similar zero results.

We also computed an index of *functional performance* (FP) for the first time with these data, measuring the range of functions for which the local government is responsible (identical to that proposed in Clark, Ferguson, & Shapiro, 1982, except it includes both common and non-common functions). Cities with high FP scores perform education, health, and social service functions, as well as common functions such as police and fire services. High-FP cities adopt more strategies of all three sorts. This may be since the broader range of functions provides more potential for innovation, generating an (interactive) synergy among staff who see more ideas being tried in their city. Cuts in intergovernmental grants are also more likely in cities with wider responsibilities, which may in turn stimulate local efforts to maintain programs with alternative strategies. Staff may also differ on other characteristics in high-FP cities.

The reform government and FP measures of governmental structure may inadequately capture an emphasis on "administration" rather than "politics" in actual policymaking. To measure *professionalization* and (perhaps) rationalization of decision making, we computed a measure of the *relative power of elected political officials versus staff*. Do powerful staff members adopt more strategies? Staff are professionally rewarded for sound fiscal policy, and might thus pursue such strategies. Professional and academic journals discuss a wide range of strategies. These provide a cafeteria array of choices for professionally informed staff.

Opposing this argument is another from organizational sociology. "Reformed" governments are bureaucracies: Officials are specialized, appointed for technical competence, roughly guaranteed tenure, and indifferent to partisan politics (Weber, 1947). Weber stresses that bureaucracies escape the control of their nominal masters, elected politicians. Yet if bureaucracy may be the technically best means to a given end, it is inflexible: Changes in organizational goals (by elected officials) may be resisted; changes in the organizational environment may be ignored (Hannan & Freeman, 1977, 1984). This bureaucratic argument suggests that reformed governments — and staff-dominated governments, regardless of legal form — will be less innovative.

To measure "rationalization" as staff dominance, we use mayors' responses about the relative influence of elected versus appointed officials.[4] Cities where staff were more important were slightly more likely to use more strategies for productivity improvement, but not other types of strategies. Still, staff power was bound up with other leadership measures such that in most regression analyses it usually fell below significance.

Next are *citizen preferences and population characteristics*. Population measures can index citizen "demand" (termed "needs" when they become politically legitimate). A more neutral term is simply *preferences*. We hypothesized, following the review of citizen surveys in Clark and Ferguson (1983, pp. 176-182), that the "nonpoor" (middle class) prefer lower taxes and spending, while the poor and blacks of all income levels prefer more public services. As operational measures, we use the percentage of the city's 1979 population above the poverty line and the percentage black (U.S. Department of Commerce, 1983). Both affect leadership patterns, but they have no direct effects on strategy adoption. Leaders' reports on citizen preferences are discussed below.

A major contribution of the FAUI Project is the collection of detailed information about many aspects of *political and administrative leadership*. These permit the study of many political processes in a way never previously possible on such a large scale. Many past studies thus had to refer to case studies, or omit such processes, or speculate about them with quite indirect indicators (e.g., using the number of employees as a rough measure of union strength). Much work on urban policy seems to be moving toward considering political processes in a more fine-grained manner, and toward study of how these processes interpenetrate other local processes. We are pleased to add new findings that contribute to a growing body of work on urban political processes in this last section.

Most results discussed thus far have been broadly similar to those of the past seven studies. However, we have recently created numerous new measures of political processes from our survey items, with dramatic results. They are so new that we and others have not yet been able to analyze them thoroughly, yet we feel compelled to signal their importance so others can help assess them. The new measures largely concern citizen preferences, organized group preferences, and the mayor and council's social background characteristics, activities, and preferences. These new measures (1) are often moderately interrelated with each other; for example, cities whose mayors prefer more spending also have council members, citizens, and organized groups who prefer more spending, as well as more active and influential organized groups. This led us to attempt to summarize their effects in summary measures that might tap a single general dimension. We thus created two multiplicative terms of several such individual indicators and measures of partisanship. (2) The new measures correlate rather neatly with four key types of political culture, discussed below. (3) Many of the new measures are positively associated with use of all three sorts of strategies (revenue, expenditure, and productivity). This led us to include many of these new measures in larger regression models. As more variables are added, the R^2s increase, sometimes to over .50 — dynamite results, undercutting some early interpretations of these data suggesting that strategies may be adopted in almost "garbage can" manner. Still, modeling problems clearly remain.[5] We thus cautiously present mainly simple correlations (Table 2.6) with the four political cultures and the three strategy types, and include in the reduced-form model only *mayors' spending*mayors' influence* and proportion of Democrats in the council.[6] We comment briefly on each finding, as these measures are so new.

One way to join the various elements of political leadership is to consider how leadership patterns cluster; the main types emerge from distinct *political cultures*. Recent work emphasizing political culture includes that of Wildavsky and Douglas (1983), Knoke (1981), and Clark and Ferguson (1983). We here extend the approach of the last work. It specifies transformation rules of five deep structures (like fiscal liberalism) that define four political cultures.

Democrats follow the agenda forged during Franklin Roosevelt's New Deal. They are fiscally liberal (committed to spending policies of the interventionist state), socially progressive (committed to racial equality, environmental concerns, and so on), and compete for office using organized groups (traditional political party clubs, unions). They concede the legitimacy of organized group interests, as well as of groups' use of politics to maximize their own self-interest. *Ethnic* or *clientelist* political culture is similar to the Democratic, but stresses more a particular ethnic base. Third, *Republican* political culture is fiscally and socially conservative, suspicious of and largely independent of organized interest groups, and committed to pursuit of collective goods — like low taxes — rather than clientelistic benefits.

A novel political culture, the *New Fiscal Populist* (NFP), emerged in the 1970s campaigns of politicians like Pittsburgh Mayor Peter Flaherty, San Francisco's Dianne Feinstein, and U.S. Senator Gary Hart. It contains several elements of Republican political culture, but replaces Republican "social conservatism" with the "social progressivism" of the Democrats. Critical of traditional organized groups (unions and parties) of the New Deal coalition, NFPs make a populist appeal to individual citizens. Not indebted to specific organized groups who seek favors, they support "public goods" as policy outputs that benefit the "whole community."

To operationalize political culture, we use mayors' responses about the five dimensions mentioned above. Each city is thus scored on the four culture indexes: Democratic, Republican, New Fiscal Populist, or

TABLE 2.6
New Political Process Measures: Correlations with Four Political Cultures and Three Sets of Strategies

	Political Cultures				Strategies		
	Democratic LDPCIX	New Fiscal Populist LNPCIX	Republican LRPCIX	Ethnic/ Clientelistic LEPCIX	Revenue LRRS	Expenditure LEXRS	Productivity LEADMP
Fiscal management index	.015	.083	.022	-.088	.056	.110**	.337**
Women mayors	-.015	.107**	-.021	-.186**	.163**	.200**	.175**
Staff power	.249**	-.225**	-.273**	.122**	.075	.054	.078
Policy distance of mayor from citizens	.204**	-.212**	-.256**	.190**	.081	.191**	.089
Policy distance of mayor from council	-.271**	.287**	.310**	-.213**	.011	-.037	-.049
Democratic council members	.104	-.096	-.110*	.222**	-.086	-.102*	-.173*
Mayor's spending preferences	.371**	-.354**	-.407**	.299**	.042	.074	.101*
Mayor's spending* influence	.357**	-.334**	-.383**	.206**	.128**	.121**	.129**
Spending preferences of organized group	.316**	-.275**	-.338**	.193**	.146**	.173**	.119**
More active organized groups	.304**	-.279**	-.351**	.226**	.149**	.194**	.084
Business group spending preferences	.093*	-.131**	-.105*	.067	.049	-.015	.071
Business group activities	.322**	-.340**	-.349**	.173**	.036	-.009	-.031
Citizen spending preferences	.210**	-.235**	-.232**	.190**	.042	.006	.037
Council member spending preferences	.106	-.116	-.135*	.090	.008	.027	-.035

Federal government-related problems	.151**	-.122*	-.156**	.197**	.171**	.195**	.089*
Nonwhite mayor	.170**	-.143**	-.181**	.217**	.042	.063	.027
Percentage black residents	.161**	-.188**	-.219**	.280**	.084*	.109**	.012
Proportion black council members	.122**	-.153**	-.134**	.138**	.006	.010	-.085*
Proportion nonwhite council members	.119**	-.165**	-.123**	.152**	-.006	.007	-.099**
Protestant mayor	-.085	.062	.075	-.204**	.040	-.049	-.029
Proportion municipal employees on council	.094*	-.144**	-.151**	.119**	.056	.025	-.018
Percentage municipal employees in bargaining units	.097*	-.055	-.084*	.071	.030	.134**	.048
Mayor's spending on municipal employees	.228**	-.259**	-.312**	.123**	.123**	.118**	.093*
Proportion independent council members	-.060	.089*	.059	-.113**	.019	.002	.025
Mayor's use of media	.144**	-.187**	-.179**	.253**	-.011	.046	.020
Sum of all problems	.053	-.026	-.057	.100	.186**	.255**	.119**
Pressure from all participants	.354**	-.336**	-.390**	.266**	.081	.053	.046
Pressure from major participants	.347**	-.313**	-.390**	.255**	.162*	.199**	.094*

SOURCE: Almost all measures are computed from the U.S. FAUI Project survey. Percentage organized municipal employees is from the 1982 U.S. Census of Governments; percentage black is from the 1980 Census of Population (included here for comparison with the black political measures).

NOTE: This table presents simple correlations (Pearson r's) indicating generally significant relations among 28 new political process measures, the 4 political cultures, and 3 sets of strategies. Bivariate Ns vary from .213 to .516.

*Significant at .05 level; **significant at .01 level, one-tailed test. Some results discussed in the text are significant at slightly lower levels.

55

Ethnic.[7] Note that Democratic and Republican political cultures are not defined here by party affiliation, but by the five general dimensions. The political cultures have often explained more about urban policy than party identification.

The four political cultures and the three types of strategies are correlated with the many "new" measures in Table 2.6. Consider specific results. *Sophisticated fiscal management* techniques[8] are more common in cities with New Fiscal Populist political cultures, and use of such techniques substantially increases use of productivity strategies. This is one of the strongest and most consistent findings in the correlations and regressions by ourselves and others in the United States and European countries that have analyzed this measure (e.g., the Norwegian study in this volume). It nicely matches our expectations about strategies NFPs should follow: They seek to reconcile their fiscal conservatism with their social progressivism by improving productivity. Better management tools are means for them to do more with less.

Women mayors are more often NFPs, and use more strategies of all sorts.

Staff power is higher in Democratic and Ethnic cities, and, as discussed above, increases productivity strategies.

Policy distance of the mayor from citizens (subtracting the mayor's preference for more spending from his or her estimate of citizen preferences for each of 13 policy areas, absolute value used) is higher in Democratic and Ethnic cities, as we expect from our characterization of their political cultures as more group and less citizen responsive. It also leads to use of more strategies. *Policy distance of mayor from the council* (again measured across 13 policy areas) shows the same relations with the political cultures, but is unrelated to the strategies.

Democratic council members (measured by party membership of individual council members, irrespective of whether their party was listed by their name on the ballot) are (slightly) more common in cities with Democratic and Ethnic political cultures. And they adopt fewer strategies. Recall that the political cultures were measured by more general dimensions than party identification, and we see here that the intercorrelations of the two are only modest. Political culture is thus clearly distinct from party identification.

Fiscally liberal mayors are more common in Democratic and Ethnic cities and adopt more strategies. These results are stronger if we use a measure multiplying *mayor's spending preferences times his or her power* (LMYSPINF).

Spending preferences of organized groups and citizens are higher in Democratic and Ethnic cities and lead to more strategies.[9]

Business group spending preferences (i.e., whether business groups prefer more, less, or same spending by the city government) correlate with the political cultures and specifically with spending preferences of other organized groups in the same city (results not shown), the same as Williams and Zimmermann (1981) found — undermining the simple view that business leaders are consistently fiscally conservative.

Business group activities are higher in Democratic and Ethnic cities, but lower in NFP and Republican cities — a pattern shared with other organized groups. They increase revenue and expenditure, but not productivity strategies, a perhaps surprising result that might be usefully reexamined by individual strategy (like contracting with the private sector).

Citizen spending preferences, as reported by the mayor, are higher in Democratic and Ethnic cities, but unrelated to the strategies.

Federal government-related problems are reported by chief administrative officers as more acute in Democratic and Ethnic cities; these same cities adopted more strategies. (See Table 2.1)

Council member spending preferences are higher in Democratic and Ethnic cities, but are not related to strategy adoption.

Nonwhite mayors are more common in Democratic and Ethnic cities; such mayors had no effect on strategy adoption. The same patterns hold for the proportion of *nonwhite council members*, although they and *black council members* adopt fewer productivity strategies. Cities with a high *percentage of black residents* do adopt more revenue and spending strategies (although these results fall below significance in regressions).

Protestant mayors are more common in NFP and Republican cities, but adopt no more strategies (with due respect to Max Weber).

Larger proportions of *municipal employees serving as council members* are more common in Democratic and Ethnic cities, and adopt more revenue-raising strategies, but they are unrelated to expenditure-reduction or productivity-improvement strategies (i.e., they are not negatively related to productivity, as a conservative "union" image might suggest). The same general pattern holds for the *percentage of municipal employees organized in collective bargaining units*.

Mayors who support more spending on municipal employee salaries, multiplied by the power of municipal employees, are far more common in Democratic and Ethnic cities, and are especially likely to adopt revenue and expenditure strategies.

The proportion of *independent (non-party-affiliated) council members* is higher in NFP and lower in Democratic and Ethnic cities, supporting the interpretation that NFP leaders have more trouble emerging in more partisan settings (see Clark with Balme, Becquart-Leclercq, Nevers, & Hoffmann-Martinot, 1987). But independent council members are unrelated to strategy adoption.

Mayors who use the media more heavily are more common in Democratic and Ethnic cities (yet not, to our surprise, in NFP cities), but mayors' media use is unrelated to strategy adoption.

Three indexes that sum the above items show the same patterns as most political process items. The *sum of all problems* mentioned by CAOs (reported individually in Table 2.1) correlates positively with all three strategies. *Pressure from all participants* is a multiplicative interaction term of participants' spending preferences times participants' activities.[9] *Pressure from major participants* is similar to the all-participants measure, but omits federal and local government officials. The major-participants measure is more strongly related to the strategies. A "state-centered" interpretation is weakened by the results that the measure including the preferences and influence of federal and local officials has less impact on strategy selection than the measure omitting these officials. Both measures heighten the effects of spending in the correlations, and we included them in some regressions. They are associated with the mayor spending*power interaction term, and are sometimes suppressed by it. They probably should be modeled as causes of mayor's preferences in a path-recursive model.

Clearly, more precise associations, causes, and consequences of these new measures must be specified to integrate them into a more comprehensive analysis.

Conclusions

On many key points, we find patterns conflicting with common interpretations of American cities. This is largely due to common reliance on national aggregate data; instead, we draw on analysis of individual cities.

Starting with reports by city chief administrative officers on major problems facing cities, we find that they stress extralocal factors such as inflation. Turning to the strategies adopted from a list of 33, many are being used. But most American cities are not cutting back substantially on their functions or work forces; instead, they are looking for cheaper ways or alternative funding to do what they already do. The most popular strategies seem to be the least threatening to established bureaucratic interests and offer tangible benefits to some staff. User fees are a case in point: Their adoption avoids cutbacks, necessitates no changes in agency procedures, and may increase autonomy of the agency adopting the fee. User fees are the most common of the 33 strategies pursued by American cities.

What leads cities to use different strategies? Surprisingly weak are characteristics of the local government's environment (e.g., tax base, population composition) or fiscal strain (measured by changes in revenue or spending, and spending divided by the tax base). These "non-findings" have important policy implications. Many public officials discussing strategies they might adopt often comment that "we are too poor" to do X. Or "federal cuts caused" us to do Y. Yet our findings suggest that such economic arguments are usually unfounded. Political leaders and staff — their policy preferences, responsiveness to different organized groups, and relative powers — are the key explanations of different strategies. Our rich data file has permitted us to measure many such processes, which we are pleased to be able to add to the growing body of work on the more fine-grained analysis of urban political and administrative processes and how they affect specific policy outputs.

Appendix:
Replication of Norwegian and French Analyses

We could replicate almost exactly the Norwegian regression, and most French regressions explaining the strategies. Our replication yields results remarkably similar to findings for Norway. Local wealth is per capita income 1979; local political ideology is proportion Democratic council members; mayoral spending and administrative sophistication match the Norwegian items; intergovernmental transfers include federal and state monies; fiscal stress is measured in two ways that are both insignificant — common function expenditures 1980/per capita income and own source revenue in 1984/own source revenue 1980. To use just tax revenues as the Norwegians did would introduce noncomparability across jurisdictions, since U.S. municipalities vary in revenue structure. We thus used revenues from all local sources.

Richard Balme reports three French equations in this volume using the strategies as dependent variables. We replicated these with U.S. data and found that many patterns differed. Some differences are no doubt due to our not using directly comparable measures. The U.S. results (replicating Balme's equations 1, 4, and 10) are presented below. The French notation is replicated, showing

TABLE 2.A1

Independent Variables	Revenue Index	Dependent Variables	
		Expenditure Index	Productivity/ Management Index
Population size	.156*	.187*	.054
Local wealth	.024	−.049	.025
Local political ideology	−.110	−.152*	−.164**
Mayoral spending preferences	.029	.053	.076
Administrative sophistication	.017	.064	.304**
Intergovernmental transfers	.034	.025	.031
Fiscal stress	.024	.084	.079
R^2	.043	.084	.155
Adjusted R^2	.005	.049	.122

*Significant at .05 level; ** significant at .01 level.

beta coefficients and significance levels in parentheses. DF = degrees of freedom:

$$LEXRS = .050LMYSPSUM + .147LV799 + .033LNWHI80 - .076LTD80PCI$$
$$\quad\quad (.37) \quad\quad\quad (.02) \quad\quad\quad (.57) \quad\quad\quad\quad (.18)$$
$$\quad + .164LFPIDX84$$
$$\quad\quad (.007)$$
$$R^2 = .073 \quad DF = 300 \quad\quad SigF = .0003 \quad\quad\quad\quad [1]$$

where LEXRS = strategies to reduce expenditures; LMYSPSUM = sum of mayor's spending preferences; LV799 = population size; and LNWHI80 = non-white population in 1980, which seems the best analogy to French foreign measure, as in France they are largely North African workers. Level used as Balme reports that level and change generate similar results; LTD80PCI = fiscal strain as function of total debt/per capita income; LFPIDX84 = index of functional performance, included to permit comparison of debt across municipalities, since debt is a noncommon function in the United States; it is similarly included in the next two equations, since they use another noncommon function: general expenditures. The only similar result is that expenditure strategies are more common in larger cities in both countries.

$$LEXRS = .169LPV897 - .114LGEW80C + .046LMYSPSUM$$
$$\quad\quad (.020) \quad\quad\quad (.167) \quad\quad\quad (.409)$$
$$\quad + .092LNWHI80 + .192LUNEMP80 + .064LPOLD80$$
$$\quad\quad (.159) \quad\quad\quad (.004) \quad\quad\quad (.331)$$
$$\quad + .067LPV900 + .309LFPIDX84$$
$$\quad\quad (.298) \quad\quad\quad (.0002)$$
$$R^2 = .091 \quad DF = 297 \quad\quad SigF = .0004 \quad\quad\quad\quad [4]$$

where LEXRS = strategies to reduce expenditures and LPV897 = middle income ($20,000-$30,000 in 1980). We also tried percentage below poverty, but while it had a moderate r = .08 (significant at .02 level) with expenditure strategies, it fell below significance in the regression and was thus deleted from the model since it also correlated over .5 with the two other income measures. LGEW80C = general expenditures per capita 1980; LMYSPSUM = mayor's spending preferences; LNWHI80 = percentage nonwhite; LUNEMP80 = percentage unemployed; LPOLD80 = percentage of population over age 65; LPV900 = percentage with incomes over $50,000 per capita income; LFPIDX84 = index of functional performance to adjust general expenditures. We find most variables insignificant, as in France; still, cities with more unemployment adopt more expenditure strategies, as in France; and cities with larger percentages of middle-income residents and that perform more functions adopt more strategies.

$$\text{LEADMP} = .071\text{LPV897} - .070\text{LGEW80C} + .080\text{LSTAFPW2} + .226\text{LFPIDX84}$$

$$(.225) \qquad (.388) \qquad (.157)$$
$$(.005)$$
$$R^2 = .038 \qquad \text{DF} = 302 \qquad \text{SigF} = .017 \qquad [10]$$

where LEADMP = productivity improvement strategies; LPV897 = percentage middle income; LGEW80C = general expenditures per capita; LSTAFPW2 = staff power; LFPIDX84 = index of functional performance. Here the striking result is the national difference: Productivity is more likely in U.S. cities with more powerful staffs, while in France the opposite is true. This seems to reflect the generally greater power and rigidity of French administrators compared to their U.S. counterparts — consistent with many past observations about the two countries.

Comparisons of recent French and U.S. local leadership, using several case studies as well as the FAUI Project survey, are found in Balme et al. (1987) and Clark, Hoffman-Martinot, Nevers, and Becquart-Leclercq (1987).

Notes

This is Research Report #207 and USA #88-7 of the Fiscal Austerity and Urban Innovation Project. The Inter-University Consortium for Political and Social Research, University of Michigan, Ann Arbor, is making the Project data available for general distribution. Support was generously provided by the Ford Foundation. We are grateful to Richard Balme, Rowan Miranda, Daniel Crane, Jason Tung, and Joie Bentrez for help with data preparation and analysis. Thanks go in particular to the 26 U.S. teams that collected these data and have made them available for others. Persons active in data collection and/or subsequent Project activities include the following:

Arizona: Albert K. Karnig

California: Mark Baldasarre, R. Browning, James Danzinger, Roger Kemp, John J. Kirlin, Anthony Pascal, Alan Saltzstein, David Tabb, Herman Turk, Lynne G. Zucker

Colorado: Susan E. Clarke

Washington, D.C.: Jeff Grady, Richard Higgins, Charles H. Levine

Florida: James Ammons, Lynn Appleton, Thomas Lynch, Susan MacManus

Georgia: Frank Thompson

Illinois: James L. Chan, Terry Nichols Clark, Burt Ditkowsky, Warren Jones, Lucinda Kasperson, Tom Smith, George Tolley, Laura Vertz, Norman Walzer

Indiana: David A. Caputo, David Knoke, Michael LaWell, Elinor Ostrom, Roger B. Parks, Ernest Rueter

Kansas: Robert Lineberry, Paul Schumaker

Louisiana: W. Bartley Hildreth, Robert Whelan

Maine: Lincoln H. Clark, Khi V. Thai

Maryland: John Gist

Massachusetts: Dale Rogers Marshall, Peter H. Rossi, James Vanecko

Michigan: William H. Frey, Bryan Jones

Minnesota: Jeffrey Broadbent, Joseph Galaskiewicz

Nebraska: Susan Welch

New Hampshire: Sally Ward

New Jersey: Jack Rabin, Joanna Regulska, Carl Van Horn

New York: Roy Bahl, Paul Eberts, Esther Fuchs, John Logan, Melvin Mister, Robert Shapiro, Joseph Zimmerman

North Carolina: John Kasarda, Peter Marsden

Ohio: Steven Brooks, Jesse Marquette, Penny Marquette, William Pammer

Oklahoma: David R. Morgan

Oregon: Bryan Downes, Kenneth Wong

Pennsylvania: Patrick Larkey, Henry Teune, Wilhelm van Vliet

Puerto Rico: Carlos Muñoz

Rhode Island: Thomas Anton, Michael Rich

Tennessee: Mike Fitzgerald, William Lyons

Texas: Charles Boswell, Richard Cole, Bryan Jones, Mark Rosentraub, Robert Stein, Del Tabel

Virginia: Robert DeVoursney, Pat Edwards, Timothy O'Rourke

Washington: Betty Jane Narver

Wisconsin: Lynne-Louise Bernier, Richard Bingham, Brett Hawkins, Robert A. Magill

Wyoming: Cal Clark, Oliver Walter

1. The revenue totals for all U.S. cities in Table 2.2 include utility and insurance trust revenues, although these are often reported separately since they are often administered semiseparately from the local government. Intergovernmental grants appear smaller than if utilities were excluded. However, utilities are excluded from the Figure 2.1 base of own source revenues.

2. The question format appears in the note to Table 2.3. Our classification of revenue enhancement, expenditure reduction, and productivity improvement strategies was operationalized in seven indexes using different specific combinations. The main three are nine revenue strategies, twenty expenditure strategies, and three productivity strategies

("decrease administrative expenses" plus the two "productivity" items). Responses were coded from 5 for "very important" to 1 for "don't know/not applicable." For regression analyses we wanted to preserve as many cases as possible, but many CAO respondents left items blank rather than checking "DK/NA" for those strategies they were not using. Based on several discussions with informants and researchers, we have generally recoded such "nonresponses" to the very low score of 0.5 (rather than "missing"). Still, if a CAO checked no strategies at all, we coded the case as missing. Each city for which the CAO reported that some strategies were being used thus has a score on each of the 33 strategies ranging from 5 to 0.5. These scores for individual items are summed for each city in our indexes of revenue raising, expenditure reduction, and productivity, which are the dependent variables in the regressions. We are examining, and encourage others to help explore, alternative coding decisions for their effects. Details on recoding, index construction, and the like are reported in Bickford, Clark, and Ritchie (1987, see especially Addenda I and II). The data are available with indexes and recodes documented in files in the "Computes" directory.

3. This background led us to specify the "structural model"

$$S_i = aT^{b1}F^{b2}G^{b3}A^{b4}C^{b5}O^{b6}L^{b7}e$$

where S_i = use of strategy i (where S_is are LRRS = revenue, LEXRS = expenditure, LEADMP = productivity strategy); T = tax base; F = fiscal strain; G = government structure; A = administrative staff power; C = citizen preferences; O = organized group activities; L = leader preferences; a = constant; b1 to b7 = coefficients to be estimated; and e = error term.

The multiplicative functional form, as suggested by Cobb and Douglas, is used since the explanatory variables interact; for example, higher fiscal strain and a weaker tax base should both individually and jointly create more environmental pressure on leaders to generate new coping strategies. The equation was estimated by taking logs of both explanatory and dependent variables, and applying OLS regression. Several variables were insignificant in the structural equation, so we removed them and specified a "reduced-form" model, shown in Table 2.5.

Census population and fiscal data are available for 1,030 cities with populations over 25,000 in 1970 or 1980. Surveys were sent to mayors, CAOs, and the heads of council finance committees, about half of whom responded, but not to every item. Thus Ns vary from a low of 213 to 1,030. Regression analyses were estimated using both listwise and pairwise treatment of missing cases. Compared to past urban research with original political data, this large N reduces multicollinearity and permits use of more independent variables.

4. Responses were combined from three similar items posed to the mayor: (1) "In the last three years, how important was the professional staff as compared to elected officials in affecting the overall spending level of your city government?" (2) "How about in allocating funds among departments?" (3) "How about in developing new fiscal management strategies, such as imposing user charges for swimming or contracting out for services like garbage collection?" Response categories ranged from 1, "Professional staff largely set level," to 5, "Elected officials largely set level" for Question 1. Instead of *level* as in the Question 1 responses, Question 2 responses substituted the word *allocations* and Question 3 responses used *decide*.

5. The new measures are sufficiently interrelated that individual coefficients grow unstable as the equation grows. Clearly, some of the higher R^2 results reflect exploding

coefficients from multicollinearity. R^2s also rise using listwise deletion of missing cases and other procedures that reduce the number of cases.

6. The r's are calculated using pairwise deletion of missing cases, with Ns varying from about 213 for council member items to 388 for mayor items. The high r's are thus not an artifact of comparing items for the cities that responded to the questionnaire out of the full N of 1,030 (cities over 25,000 population in 1970 or 1980). Some measures are similar to those defined at the 1987 Bergen Fiscal Austerity and Urban Innovation Project conference.

7. The political culture variables are constructed from mayoral responses to individual survey items that were joined in five indexes. The social conservatism index uses responses to questions about handguns, sex education, and abortion. The "populism" index uses questions about the responsiveness of government to citizens and how often leaders take positions unpopular with their constituents. An index of interest group activity sums items about how often the government responded favorably to public employees, low-income groups, homeowner groups, neighborhood groups, civic groups, minority groups, taxpayer groups, elderly groups, businesses, and church and religious groups. Public goods orientation is measured by response to a question about whether or not leaders should break some rules to help people. Finally, a scale of fiscal conservatism comes from mayors' reports of their own spending and tax preferences. Exact index construction is reported in Clark et al. (1984) and Bickford et al. (1987, pp. 62-65). Every city for which there are adequate survey data is assigned a score for each of the four political cultures. To simplify exposition, we may state, for example, that "Democratic" cities do more of a policy. Yet we literally mean that across all cities, those with higher Democratic index scores are more likely to follow the policy.

8. Our fiscal management index summed five items concerning sophistication of revenue forecasting, fiscal information systems, performance indicators, accounting and financial reporting, and economic development analysis (Bickford et al., 1987, pp. 62-65).

9. This measure and the next sum responses for 14 participants: municipal employees and their unions, low-income groups, homeowner groups, neighborhood groups, civic groups, minority groups, taxpayers' associations, business groups, local media, elderly, churches and religious groups, individual citizens, Democratic party, and Republican party.

10. To permit others to explore the leadership measures in Table 2.6, we list their acronyms from data file CORE4OL.SYS in the order in which they appear in the table: LFMIX SEXMAY LSTAFPW2 POLDISSN LPOLDSTC LDEMCMB2 LMYSPSUM LMYSPINF SPIG7 INVIG8 SPBIZ7 INFBIZ9 LCZSPSUM LCMBSPSM LFEDPROB LNONWHMY LBLK80 LBLKCMBS LNWHICMB LPROTMY LMECMBS LPCTORG LMRSME LINDCMBS LMEDIAMY PROBSALL PRESSALL IGSPACT.

References

Abney, G., & Lauth, T. P. (1986). *The politics of state and city administration.* Albany: State University of New York Press.

Anton, T. J. (1984). Intergovernmental change in the United States: An assessment of the literature. In T. C. Miller (Ed.), *Public sector performance: A conceptual turning point* (pp. 15-64). Baltimore: Johns Hopkins University Press.

Appleton, L. M. (1988). Determinants of innovation in urban fiscal strategies. In T. N. Clark, W. Lyons, & M. Fitzgerald (Eds.), *Research in urban policy* (Vol. 3). Greenwich, CT: JAI.

Balme, R., Becquart-Leclercq, J., Clark, T. N., Hoffmann-Martinot, V., & Nevers, J. Y. (1987). New mayors: France and the U.S. *Tocqueville Review, 8,* 263-278.

Benton, J. E., & Morgan, D. R. (1986). The intergovernmental dimension of public policy. In J. E. Benton & D. R. Morgan (Eds.), *Intergovernmental relations and public policy* (pp. 3-12). New York: Greenwood.

Bickford, A., Clark, T. N., & Ritchie, A. (1987). *FA for PC: Hints and help with SPSS/PC+ for participants in the Fiscal Austerity and Urban Innovation Project.* Chicago: Urban Innovation Analysis, Inc., and University of Chicago.

Browning, R. P., Marshall, D. R., & Tabb, D. H. (1984). *Protest is not enough.* Berkeley: University of California Press.

Burg, M. H. (1986). *The effect of political culture on urban financial decisions.* Unpublished master's essay, University of Chicago, Department of Political Science.

Clark, C., & Walter, B. O. (1986). City fiscal strategies. In T. N. Clark (Ed.), *Research in urban policy* (Vol. 2B, pp. 89-115). Greenwich, CT: JAI.

Clark, C., & Walter, B. O. (1987). Political culture, management style and fiscal austerity strategies. Paper presented at the annual meeting of the Midwest Political Science Association, Chicago.

Clark, C., & Walter, B. O. (in press). The decline of intergovernmental aid to U.S. cities. In T. N. Clark (Ed.), *Austerity and innovation in American cities.* Newbury Park, CA: Sage.

Clark, T. N. (Ed.). (1976, June). Citizen preferences and urban public policy [Special issue]. *Policy and Politics, 4.*

Clark, T. N. (1985). *Revenues for Chicago* (White Paper 1). Chicago: City of Chicago, Office of the Comptroller.

Clark, T. N., with Balme, R., Becquart-Leclercq, J., Nevers, J.-Y., & Hoffmann-Martinot, V. (1987). *Political culture and urban innovation.* Paper presented at the annual meeting of the American Sociological Association, Chicago.

Clark, T. N., Burg, M. H., & de Landa, M.D.V. (1984). *Urban political cultures and fiscal austerity strategies.* Paper presented at the annual meeting of the American Political Science Association, Washington, DC.

Clark, T. N., & Ferguson, L. C. (1983). *City money: Political processes, fiscal strain, and retrenchment.* New York: Columbia University Press.

Clark, T. N., Ferguson, L. C., & Shapiro, R. Y. (1982). Functional performance analysis. *Political Methodology, 8,* 187-223.

Clark, T. N., Hoffmann-Martinot, V., Nevers, J.-Y., & Becquart-Leclercq, J. (1987). *L'innovation municipale à l'épreuve de l'austerite budgetaire* (report from CERVELIEP, Bordeaux). Paris: Ministère de l'Équipement, Plan Urban.

Clegg, S., & Dunkerley, D. (1980). *Organization, class and control.* London: Routledge & Kegan Paul.

Dahl, R. (1961). *Who governs?* New Haven, CT: Yale University Press.

Downs, A. (1967). *Inside bureaucracy.* Boston: Little, Brown.

Elazar, D. (1972). *American federalism* (2nd ed.). New York: Crowell-Collier.

Eulau, H., & Prewitt, K. (1973). *Labyrinths of democracy.* Indianapolis: Bobbs-Merrill.

Hamilton, E. K. (1980). On nonconstitutional management of a constitutional problem. In C. H. Levine (Ed.), *Managing fiscal stress* (pp. 52-70). Chatham, NJ: Chatham House.

Hannan, M. T., & Freeman, J. (1977). The population ecology of organizations. *American Journal of Sociology, 82*, 929-964.

Hannan, M. T., & Freeman, J. (1984). Structural inertia and organizational change. *American Sociological Review, 49*, 149-164.

Hawkins, B. W. (1988). A comparison of local and external influences on city strategies for coping with fiscal strain. In T. N. Clark, W. Lyons, & M. Fitzgerald (Eds.), *Research in urban policy* (Vol. 3). Greenwich, CT: JAI.

Heidenheimer, A. J., Heclo, H., & Adams, C. T. (1983). *Comparative public policy: The politics of social choice in Europe and America* (2nd ed.). New York: St. Martin's.

Hickham, D., Berne, R., & Stiefel, L. (1981). Taxing over tax limits. *Public Administration Review, 41*, 445-453.

Hunter, F. (1953). *Community power structure.* Chapel Hill: University of North Carolina Press.

Janowitz, M. J. (1978). *The last half-century.* Chicago: University of Chicago Press.

Kirlin, J. J. (1984). Accommodating discontinuity: Adjusting the political system of California to Proposition 13. In C. H. Levine & I. Rubin (Eds.), *Fiscal stress and public policy* (pp. 69-88). Beverly Hills, CA: Sage.

Knoke, D. (1981). Urban political cultures. In T. N. Clark (Ed.), *Urban policy analysis: Directions for future research* (pp. 203-226). Beverly Hills, CA: Sage.

Larsen, H. O. (1987). *Political authority and local leadership.* Paper presented to the European Consortium for Political Research, Amsterdam.

Levine, C. H., & Rubin, I. (Eds.). (1980). *Fiscal stress and public policy.* Beverly Hills, CA: Sage.

Levine, C. H., Rubin, I. S., & Wolohojian, G. C. (1981). *The politics of retrenchment.* Beverly Hills, CA: Sage.

Lineberry, R. L., & Fowler, E. P. (1967). Reformism and public policies in American cities. *American Political Science Review, 61*, 701-716.

Lovell, C., & Tobin, C. (1981). The mandate issue. *Public Administration Review, 41*(3), 318-331.

Lucier, R. L. (1980). Gauging the strength and meaning of the 1978 tax revolt. In C. H. Levine (Ed.), *Managing fiscal stress* (pp. 123-138). Chatham, NJ: Chatham House.

Lynd, R. S., & Lynd, H. M. (1937). *Middletown in transition.* New York: Harcourt, Brace.

MacManus, S. A. (1978). *Revenue patterns in U.S. cities and suburbs: A comparative analysis.* New York: Praeger.

MacManus, S. A. (1983). State government: The overseer of municipal finance. In A. M. Sbragia (Ed.), *The municipal money chase: The politics of local government finance* (pp. 145-184). Boulder, CO: Westview.

Mohr, L. (1982). *Explaining organizational behavior.* San Francisco: Jossey-Bass.

Mollenkopf, J. (1983). *The contested city.* Princeton, NJ: Princeton University Press.

Morgan, D. R., & Pammer, W. J., Jr. (1986). *Coping with fiscal stress.* Paper presented at the annual meeting of the Midwest Political Science Association, Chicago.

Morgan, D. R., & Pelissero, J. P. (1980). Urban policy: Does political structure matter? *American Political Science Review, 74*, 999-1005.

Mouritzen, P. E., & Nielsen, K. H. (1988). *Handbook of urban fiscal data.* Odense, Denmark: Odense University.

Mushkin, S. J., & Vehorn, C. L. (1980). User fees and charges. In C. H. Levine (Ed.), *Managing fiscal stress* (pp. 222-234). Chatham, NJ: Chatham House.

Nathan, R. P., & Doolittle, F. (1983). *The consequences of cuts.* Princeton, NJ: Princeton University Press.

Pammer, W. J. (1986). *Cutback management in American cities.* Unpublished doctoral thesis, University of Oklahoma, Department of Political Science.

Peterson, P. E. (1981). *City limits.* Chicago: University of Chicago Press.

Pfeffer, J., & Salancik, G. R. (1978). *The external control of organization.* New York: Harper & Row.

Pfiffner, J. P. (1983). Inflexible budgets, fiscal stress, and the tax revolt. In A. M. Sbragia (Ed.), *The municipal money chase: The politics of local government finance* (pp. 37-66). Boulder, CO: Westview.

Rigos, P. N. (1986). *The role of user fees in municipal budgets: Fiscal health tool or spending device?* Paper presented at the annual meetings of the Southwest Political Science Association, San Antonio, TX.

Rubin, I. S. (1982). *Running in the red.* Albany: State University of New York Press.

Sbragia, A. M. (1983a). The 1970s: A decade of change in local government finance. In A. M. Sbragia (Ed.), *The municipal money chase: The politics of local government finance* (pp. 9-36). Boulder, CO: Westview.

Sbragia, A. M. (1983b). Politics, local government, and the municipal bond market. In A. M. Sbragia (Ed.), *The municipal money chase: The politics of local government finance* (pp. 67-112). Boulder, CO: Westview.

Schick, A. (1980). Budgetary adaptations to resource scarcity. In C. H. Levine & I. Rubin (Eds.), *Fiscal stress and public policy* (pp. 113-134). Beverly Hills, CA: Sage.

Scott, W. R. (1981). *Organizations: Rational, natural and open systems.* Englewood Cliffs, NJ: Prentice-Hall.

Skocpol, T. (1985). Bringing the state back in: Strategies of analysis in current research. In P. B. Evans, D. Rueschemeyer, & T. Skocpol, *Bringing the state back in* (pp. 3-43). Cambridge: Cambridge University Press.

Stein, R. M. (1982). The political economy of municipal functional responsibility. *Social Science Quarterly, 3,* 530-548.

Stein, R. M. (1984). Municipal public employment: An examination of intergovernmental influences. *American Journal of Political Science, 28,* 636-654.

Stein, R. M., Sinclair, E. G., & Neiman, M. (1986). Local government and fiscal stress: An exploration into spending and public employment decisions. In M. Gottdiener (Ed.), *Cities in stress: A new look at the urban crisis* (pp. 111-143). Beverly Hills, CA: Sage.

Stone, C. (1980). Systemic power in community decision making. *American Political Science Review, 74,* 978-990.

Tabb, W. K. (1984). Urban development and regional restructuring: An overview. In L. Sawers & W. K. Tabb (Eds.), *Sunbelt/Snowbelt: Urban development and regional restructuring* (pp. 3-18). New York: Oxford University Press.

Thoenig, J.-C. (1986). Master plan and system effects. In T. N. Clark (Ed.), *Research in urban policy* (Vol. 2A, pp. 77-88). Greenwich, CT: JAI.

Truman, D. (1951). *The governmental process.* New York: Alfred A. Knopf.

U.S. Department of Commerce, Bureau of the Census. (1983). *City government finances in 1981-1982.* Washington, DC: Government Printing Office.

U.S. Department of Commerce, Bureau of the Census. (1986). *City government finances in 1969-70, 1974-75, 1979-80, 1984-85.* Washington, DC: Government Printing Office.

U.S. Congress, Advisory Committee on Intergovernmental Relations. (1979). *State-local finances in recession and inflation: An economic analysis.* Washington, DC: Government Printing Office.

Verba, S., & Nie, N. (1972). *Participation in America: Political democracy and social equality.* New York: Harper & Row.

Weber, M. (1947). *The theory of social and economic organization* (A. H. Henderson & T. Parsons, Eds.). Glencoe, IL: Free Press.

Wildavsky, A. (1964). *The politics of the budgetary press.* Boston: Little, Brown.

Wildavsky, A. P., & Douglas, M. (1983). *Risk and culture.* Berkeley: University of California Press.

Williams, J. A., & Zimmermann, E. (1981). American business organizations and redistributive preferences. *Urban Affairs Quarterly, 16,* 453-464.

Yates, D. (1977). *The ungovernable city: The politics of urban problems and policymaking.* Cambridge: MIT Press.

3

Free to Choose?
The Case of Affluent
Norwegian Municipalities

HARALD BALDERSHEIM
SISSEL HØVIK
HELGE O. LARSEN
LAWRENCE ROSE
NILS AARSÆTHER

During the 1970s, Norway became an oil-producing country. Both directly and indirectly, exploitation of North Sea oil resources had positive consequences for public finance — directly through the taxation of oil companies, and indirectly through the taxation of increased personal incomes.

During the same period, implementation of ambitious welfare society programs were increasingly made the responsibility of local government in Norway. Welfare state goals in Norway are partly a question of service levels and partly a question of geographic access. Geographical conditions in particular represent a special challenge for service delivery. Of the country's 454 primary municipalities, 251 have fewer than 5,000 inhabitants, and many of the smaller ones have their populations scattered in communities located on islands, in communities without road connections to the municipal center, or simply too far from the center of the municipality to be served by just one school or one doctor's office alone. The idea of locating public service institutions at the lowest workable geographical level is therefore a political ambition with obvious financial implications.

The decision to use local municipalities as the primary vehicle for a policy of geographical equalization in Norway led to a major increase in municipal expenses and investments from the mid-1960s onward. It is important to stress in this connection that at a time when most Western European countries were hit by a recession due to effects of the oil crisis in 1973, Norway found itself in a twofold situation of prosperity: Not only had oil resources become available for extraction, but the expected value of these resources had increased enormously in a short period of time.

The result, which distinguishes Norway from other Western European countries, was that a period of local government expansion was prolonged and lasted until about 1980. One of the paradoxes of this growth period, moreover, was that central government money, which was transferred to the municipalities with the explicit aim of reducing accumulated debts, was instead used to finance new investment projects. The country's decentralized demographic structure, combined with strong political support for policies of geographical equalization and public expectations concerning North Sea oil revenues, in short, led to a massive investment wave in the 1970s and eventually contributed to an overheated economy at the national level.

As a consequence of these developments, the central government finally had to put an end to municipal expansionism. This was accomplished by a combined strategy of reduced investment quotas, high interest rates (especially those levied by the state-run bank for municipalities), marginal reductions in the local tax rate, and some increase in the level of mandatory transfer payments made by municipalities to the central treasury (Hansen, 1986b, p. 141).

In this regard, 1978 was a clear turning point. Thus by 1980 the majority of Norwegian municipalities were experiencing a kind of fiscal squeeze, a squeeze due to increased service expectations from the citizens on the one hand and increasing investment costs on the other. Municipalities that in particular felt this squeeze were typically those that had ventured into a series of new projects during the 1970s, with investment costs subsidized in large part by means of various central government welfare programs. By contrast, municipalities that had held lower welfare ambitions, or at least had been more restrained in their investment programs, were less affected by the squeeze and were, in an economic sense, better off.[1]

There was no growth in ordinary municipal revenues at the time that could compensate for increases in interest costs and loan repayment expenses. Normally these conditions would be expected to lead to a classical fiscal crisis in local government. But once again the central government, after initially having curbed the investment propensities of the municipalities, stepped in as a beneficiary, increasing transfer payments back to the municipalities. Increases in central government transfer payments, in fact, more than compensated for a relative decline in local government revenues derived from personal income tax sources (the single most important revenue source for local authorities) in the 1975-1985 period.

There were, to be sure, large variations among local authorities with respect to their relative income levels and the degree to which they, subjectively at least, felt fiscally squeezed. But no authorities experienced situations where they had to lay off personnel or close down whole service departments. The problem of adaptation was more one of how to achieve a mild redistribution of resources among services in order to meet new demands — for instance, more services for the elderly — rather than a desperate struggle to maintain an existing range of services.

The questions this chapter will address, therefore, are as follows: (1) What kinds of problems do local leaders see in this situation of relative affluence? (2) What are the strategic choices they make and what are the instruments they use when trying to deal with these problems? (3) What are the factors that influence the strategic choices that are made when redistribution is necessary under conditions of nominal public affluence? Before turning to these questions, however, we will briefly look at central-local relations in Norway and major features in the development of local finances from 1975 to 1985.

Central-Local Relations in Norway and the Development of Local Finances, 1975-1985

In 1987, Norwegian municipalities feted themselves; 150 years had passed since passage of legislation establishing local self-government in Norway. With the king and prime minister in attendance, the Association of Norwegian Municipalities tried to turn the anniversary into a manifestation of the values and virtues of local self-government. Yet one may nevertheless ask just what justifications there are for speaking of local autonomy in Norway.

Local government in Norway was not initially set forth in the Constitution of 1814. Local power is instead a subsequent derivation adopted by central authorities. The original legal basis for local government was the Act of Aldermen (*Formannskapslov*) passed in 1837. Fundamental to this law were principles of lay rule and proportional representation. The act prescribed the election of a local council of representatives, from which a smaller board of aldermen (i.e., an executive committee) was to be chosen by the representatives themselves. One of the aldermen was also to be elected as mayor by the council.

By virtue of this law, a structure of local government of dual character was established in Norway: On the one hand, it is an expression and instrument of local autonomy and self-governance; on the other, it is simply the lowest level in a three-tier system of unitary political and administrative power. This dual character of local authorities – to express and transform the needs of a territorially defined population and to implement nationally mandated policies – has remained essentially the same ever since. From a comparative perspective, however, it is important to note that the discretionary authority of local government in Norway is in principle only negatively limited by the law. Municipalities may engage in any activities not explicitly forbidden to them or assigned to other units of government.

Potentially, therefore, the discretionary authority of Norwegian municipalities is substantial: So long as the municipalities comply with all relevant laws and satisfy all duties imposed upon them by central authorities, they are free to take on any task they choose. In practice, of course, this discretionary authority is limited by various considerations, fiscal solvency being prominent among them. Of particular relevance in this connection is the fact that the level of local personal income taxation in Norway is for all practical purposes outside the area of local control; that is, central authorities establish a narrow range of permissible tax rates with minimum and maximum levels, and within these limits virtually all Norwegian municipalities have adopted the same level of taxation – the maximum possible.

The Reform Process:
Toward Increased Local Self-Government?

One may reasonably argue that Norwegian municipalities, as they exist today, owe their standing to the welfare state. To a large extent the implementation of national policies within various sectors has been carried out through the municipalities rather than through separate, functional authorities. A frequently heard complaint from local councilors during the last ten years, in fact, has been that the very number of national tasks and standards to be implemented by the municipalities has made the notion of local autonomy something of a joke.

Such complaints definitely have some substance to them, but are not always entirely justified. Given the time lag between the point when investment decisions are initially made and when their consequences — particularly those involving capital expenditures and/or operating expenses — appear on the budget, local politicians have tended to ignore their own role in determining the municipality's situation and have instead placed the blame for local financial hardships on state policies in general.

Reforms of local government and central-local relations have occupied a prominent place on the political agenda in Norway for the entire postwar period. The overarching goals of this reform process have been to achieve decentralization and democratization together with a practical and effective form of administration.[2] To obtain these objectives, the municipalities had to be made more effective units for service production and delivery, which in turn implied that they had to be "economically viable units."

According to some observers, local government reform in Norway has been of a more piecemeal character than in neighboring Scandinavian countries (e.g., Kjellberg, 1981b, 1985). One may nonetheless speak of phases in the development of intergovernmental relations in Norway. The 1960s, for example, were marked by efforts to make the municipalities larger and more effective units of government, and by a belief in the importance of comprehensive planning and rational action. Hence foremost on the agenda were the amalgamation of municipalities and the introduction of new and ambitious planning programs to be carried out at all levels of government. The number of municipalities was reduced in this connection from 744 to 451 between 1960 and 1970 (a few municipal "divorces" subsequently increased this number to 454). As noted earlier, however, a majority of the municipalities still remain comparatively small by international standards.

The 1970s, by contrast, was a decade marked by implementation of welfare state programs. The municipalities experienced a period of strong economic growth as a result. Both investments and the number of municipal employees soared, until at last the state tried to apply some brakes toward the end of the decade. Yet during this same period local politicians increasingly expressed dissatisfaction with respect to the degree of central control, especially when the financial consequences of investments started to have an impact on municipal budgets. Put quite bluntly, local politicians often felt like puppets on the end of strings being pulled by the central government and thought they had been reduced to local administrators of the welfare state.

The 1980s may represent a break with this situation, at least if the rhetoric voiced in connection with the introduction of several new reform measures proves to be true. It has been argued from various perspectives that the 1980s should become a decade of organizational reform within the municipalities, and the Norwegian Association of Municipalities launched a research program to look into the issue. Of the reforms introduced so far, perhaps the most significant concerns the restructuring of the system of local standing committees. Even the smallest municipalities characteristically had a large number of such committees previously, each taking care of a small area of functional responsibility. Increasingly, however, the municipalities are adopting a system of only five or so primary committees corresponding to the principal areas of municipal expenditure.

More conspicuous than this reform, however, have been the reforms in central-local relations. This has been evident particularly with respect to municipal finance and practices regarding state transfer payments. Although considerable variation has existed among municipalities, state transfers have in general accounted for more than a third of all municipal income in Norway in recent years. (Details on municipal finance are to be found in the next section.) Until 1986 the municipalities relied upon a plethora of specialized grants in this regard. But in 1986 this system was replaced by a new block grant system establishing four sector-specific and one general grant for each municipality (see St. meld. nr. 26, 1983-84, and Ot. prp. nr. 48, 1984-85). By reducing state tendencies to engage in economic micromanagement of municipal activity, it has been argued that this reform entails genuine decentralization of responsibility and serves to increase local self-government.[3]

In contrast to this reform of the system of transfer payments, which focuses on economic management, a more recently inaugurated reform experiment is intended to investigate the consequences of lessening central control usually imposed on the municipalities through laws, regulations, and instructions. By virtue of this experiment, a few municipalities (20 of 454) have for a four-year period been granted the opportunity to seek exemption from rules that might otherwise hamper the most efficient and coordinated use of public resources. The principal thrust of this experiment is to give these municipalities greater freedom to make decisions affecting local living conditions at the municipal level, thereby gaining experience on which future decisions regarding issues of local autonomy can be made (Ot. prp. nr. 36, 1985-86; Kommunal- og arbeidsdepartementet, 1986).

TABLE 3.1

Annual Percentage Change in Volume of Gross Expenditures for
All Norwegian municipalities, 1975-1985

1975	1976	1977	1978	1979	1980	1981	1982	1983	1984	1985
+6.8	+9.1	+7.2	+7.1	+4.1	+6.1	+3.2	+5.5	+4.4	+3.5	+4.4

SOURCE: Norges offisielle statistikk (1987, pp. 218-219).

On balance, then, one may conclude that the major obstacles to innovation and policy change at the local level do not stem from constitutional or judicial constraints placed upon the municipalities. Norwegian municipalities enjoy the formal freedom to take on almost any task they choose, subject in large measure only to the constraints of resources available to them. Until the early 1960s, this prospect was in fact a rather void and empty freedom for most municipalities, because they were very small and lacked the necessary resource base. Subsequent amalgamation and implementation of welfare state policies produced considerable economic growth, but streamlining and the imposition of national standards prevented a corresponding enhancement of local autonomy. Thus, until recently, the dominant pattern of what Norwegian municipalities were doing was largely dependent upon the character of national programs that were to be implemented.

Yet, as we have indicated, it would be wrong to think that no discretionary authority or financial slack whatsoever existed previously for the municipalities. This is evident from an examination of the development of local finances in the period from 1975 to 1985. As will be seen, municipalities found means to increase investments and expenditures, and the state eventually found it difficult to put the brakes on what had become the flywheel of municipal expansion.

The Development of Local Finances, 1975-1985

The municipal sector in Norway can hardly be said to have been hit by recession in any objective sense in the years from 1975 to 1985. This is evident from the figures presented in Table 3.1, which show the annual percentage growth rate in municipal expenditures for the entire period. Although the annual growth rate reflected in Table 3.1 is generally high throughout the period, there is a drop from a rate of 7% plus in the years from 1975 to 1978 to a somewhat lower growth rate of roughly 4% in the 1980s. Such a drop may subjectively be experienced as a setback, but this can hardly qualify as a situation of fiscal austerity.

TABLE 3.2

Local Tax Income and State Transfer Revenues as a Percentage of
Municipal Expenses (excepting investments, municipal business,
and loan repayments), 1975-1985

Revenue Source	1975	1980	1985
Taxes (mainly personal income)	75.6	62.0	56.9
General state transfers	4.4	4.3	4.9
Categorical state transfers	16.5	23.0	30.5
Other incomes	3.5	10.7	7.7
Total expenses (in billion NOK)	23.3	28.6	53.1

SOURCE: Norges offisielle statistikk (1977, pp. 8.1-8.5x; 1982, pp. 8.1-8.8x; 1986a, pp. 42-47).

Behind these aggregate figures lie a host of intermunicipal variances, changes in state-local transfers, changes in local revenue structure, and differences in how money is spent within different areas of municipal activity. We will start with an examination of the structure of municipal income in the period.

Until the introduction of the new grant system in 1986, municipal income in Norway was derived from four primary sources: (1) local taxes, (2) general state transfers, (3) categorical state transfers, and (4) assorted other revenue sources. Local taxes consist first and foremost of personal income taxes. As noted previously, the central government sets the tax rate (currently about 14% of taxable net income), which is common to all municipalities. (In addition, the county collects 7% of taxable personal income to finance activities such as hospitals, secondary education, and certain transportation facilities.) Local taxes also include property taxes, but this kind of taxation is not compulsory. As a source of income, property taxes have traditionally been little used, and even though an increasing number of municipalities have begun collecting them, they continue to contribute very little to the overall municipal revenue structure.

As for other revenue sources, the municipalities may impose both income and capital asset taxes on businesses, but only on the net taxable value. Since most businesses are capable of writing off or deducting many expenses, however, this is also a generally minor source of revenue for most municipalities. Other types of business taxes are virtually nonexistent. Assorted user fees, by contrast, are more prevalent, especially for municipal utilities, but again the rates charged have traditionally been so low that these have not constituted

TABLE 3.3

Investment and Capital Expenditures in Norwegian Municipalities as a
Percentage of Total Expenses (excepting investments, business,
and loan repayments), 1975-1985

Expenditure	1975	1980	1985
Investments	20.7	19.3	11.7
Interest and repayments	8.6	16.8	17.3
Total	29.3	37.1	29.0

SOURCE: See Table 3.2.

a major source of income for the average municipality. These considerations are evident in Table 3.2.

The most interesting tendency to be observed in Table 3.2, however, is of a different variety. When local taxation and general transfers together fell behind in terms of financing ongoing municipal activities during the 1970s, the central government either was willing to or, alternatively, was forced to take responsibility for the municipal sector. Up until 1985, the central government's principal strategy in this regard was to expand categorical grants, while general transfers, intended to compensate for variations in local taxable income, scarcely changed their share.

The structure of local municipal expenditures depends, of course, on changes in ongoing programs, as well as changes resulting from new activities. Table 3.3 shows how investments and capital expenditures relate to the service levels in Norwegian municipalities.

In the 1970s, the investment rate in Norwegian municipalities was very high. In 1975, however, the other side of the coin was hardly visible — that is, interest costs and loan repayments accounted for only a modest share of total expenses. By 1980 the picture had changed; the high investment rate had by now led to high capital expenditures, due to accumulated debts as well as to restrictions on capital subsidies and cheap credits imposed by central government.[4] Investment rates have dropped in the 1980s, but the municipalities cannot expect to be relieved from the consequences of the expansion in the 1970s. Thus in 1985 a solid 17% of all municipal expenses consisted of interest costs and loan repayments.

Another issue is how the money was spent in the period 1975-1985. Was the growth evenly allocated across traditional sectors, or can we observe changes in priorities? In Table 3.4 we attempt to show the

development for three important age-specific policy fields: day-care
centers for children (0-6 years of age), primary education for children
(7-16 years), and services for the aged (67 years and older). For
purposes of comparison, we also show corresponding figures for
another type of local government activity — development of the
municipality's technical infrastructure.

The three age-specific fields are chosen because they reflect changes
in contextual conditions confronted by municipalities. A strong growth
in female employment, for example, affects the need for children's
day-care arrangements, even though the total number of children
throughout the country as a whole is decreasing. For the primary school
system, by contrast, a decreasing number of children means less demand, while a higher number of older people means more pressure for
pensioners' homes and services to elderly people living in their own
homes. The effect of an increased number of elderly, moreover, tends
to be strengthened because of increasing female employment, since this
means fewer housewives are available to look after older relatives.

Some important changes occurred with respect to the assignment of
responsibilities at the subnational level between 1975 and 1985, particularly regarding operation of medical institutions at the regional
level (the county municipalities). The growth in total expenditures for
the three age-specific sectors between 1975 and 1980 is for this reason
misleading. This problem does not affect the relations among the three
age-specific service fields, however, and here we see a marked tendency of growth in expenditures for services for children below school age
as well as for services to the elderly. In the last period, by contrast, the
two expanding sectors seem to stabilize, while there is a definite drop
in the primary school share of total expenditures. By comparison, the
share of total expenses for the technical sector — roads, water supply,
sewer, and so on — remained fairly stable or decreased slightly during
the 1975-85 period.

On the whole, it is fair to say that the structure of municipal activity
changed in the period from 1975 to 1985, but that the rate of change has
been lower for the 1980s than in the late 1970s. Factors that contribute
to this change in activity appear to be found in the system of state-local
financing (especially the growth in categorical grants) and environmental conditions (especially demographic changes) faced by municipalities. For at least two reasons, however, increases in interest and
loan repayment expenses should tend to stabilize the structure of municipal activities in the near future. First, constraints on investment funds
will provide fewer opportunities to develop new local institutions;

TABLE 3.4

Municipal Expenditures in Four Policy Fields as a Percentage Share of
Total Running Expenditure (excepting investments, loan repayments,
and municipal business), 1975-1985

Policy Field	1975	1980	1985
Day-care centers	1.8	4.3	5.0
Primary schools	18.7	26.3	23.0
Services for the aged	5.0	11.3	11.3
Total for three age-specific policy fields	25.5	41.9	39.3
Technical infrastructure	14.0	15.6	13.9

SOURCE: See Table 3.2

second, these expenses will limit a comfortable growth rate that would otherwise allow for allocation games with no obvious losers. Given this situation, it is reasonable to ask what problems local leaders may see. It is to this question that we now turn.

Important Problems as Perceived by Local Leaders

In 1985, the Norwegian FAUI group worked out two versions of the questionnaire, and sent them to mayors and chief administrative officers (CAOs) in all 454 municipalities in Norway. A letter of recommendation was enclosed from the Norwegian Association of Municipalities, and a response rate of over 80% was achieved in both cases.

In the tables that follow, we present the responses of mayors and chief administrators to questions designed to tap the local problems dimension. We then discuss similarities and discrepancies between political and administrative perceptions, and try to relate these to objective criteria.

Local Challenges: The Mayor's View

The questionnaire for mayors included an open-ended question asking respondents to write down the "most important tasks for your municipality in the next 2 to 3 years." Three numbered lines followed. Table 3.5 is based on a post hoc classification of a total of 862 tasks mentioned in response to this question (no more than three per mayor).

TABLE 3.5

Mayors' Identification of Important Tasks for Their Own Municipalities (N = 862)

Task Area	%	Dominant Single Tasks	%
Mandated services	42	Services for the aged	16
		Water, roads, sewage	7
Voluntary services	36	Development projects	29
		Communications	7
Municipal economy and organization	17	Monitoring expenses and incomes	7
Other task areas	5		

SOURCE: Norway FAUI Project survey, 1985 mayor file.

For purposes of presentation, tasks reported have been classified into two service categories — mandated and voluntary — as well as responses related to economic and organizational problems.[5]

The most striking finding presented in Table 3.5 is an unexpectedly strong orientation toward activities outside ordinary mandated areas of municipal responsibility. Some 36% of the tasks reported by mayors relate to activities in areas where the central government, regional authorities (e.g., the county municipality), or the private sector are normally expected to dominate. Within this voluntary service area, the mayors clearly favor development activities — that is, creation of local development funds and participatory investment schemes, provision of advice, and other forms of infrastructure support for private industries and business in general. Activities like these have traditionally been expected to result from interaction between central government agencies and private firms. As late as 1978, in fact, central government authorities expressed a dislike of municipal involvement in the economic development field (Aarsæther, 1985, p. 43).[6]

When mayors mention tasks and problems related to the standard repertoire of mandated municipal activities, one sector stands out in particular, namely, *services for the elderly*. Of the total number of tasks mentioned, 16% are of this type, while 7% of the tasks relate to ordinary technical infrastructure programs. Other generally demanding municipal responsibilities — such as primary education, day-care institutions, health and welfare services, churches, sports, and housing — are also mentioned, but by surprisingly few mayors.

A total of 17% of the answers were classified as tasks relating to improvement of the municipal economy and organization. Several themes — like planning, economic forecasting, organizational design, and political representation and conflicts — were mentioned in this connection, but the most frequently mentioned task concerned *monitoring expenses and incomes* (7% of the total). Under current municipal organization, monitoring measures are definitely an administrative responsibility. It is otherwise worthwhile noting that very few mayors mentioned tasks related to the political side of the organization, such as problems of representation or conflict levels.

It is important to bear in mind here that our question in no way directly taps the "problems" dimension generally found in the FAUI Project studies, since we asked the mayors to identify *important tasks for their municipalities* in the near future. Answers to our question may therefore be of a more "positive" nature than might otherwise be the case. We nonetheless presume that there is likely to be a close link between what may be identified as important areas for municipal activity in the time ahead and local problems. If our picture of the mayoral scene is valid, it shows a political leadership heavily bent toward development activities outside the traditional municipal sphere and leaders who have become aware of an aging resident population, but who are not especially preoccupied with internal organizational issues and financial management.

Only 7% of the tasks mentioned could be classified as involving problems of financial control. With a solid 29% score on development activities, one could be tempted to conclude that the mayors of the 1984-1985 period cared more for municipal expansion than for how relevant activities are to be financed. While this is a plausible conclusion, one can also offer an alternative, modified explanation stressing a perceived crisis in local business activities and a diminishing fiscal base. This alternative has a great deal of merit, particularly insofar as municipal finances depend heavily on a 14% personal income tax, collected directly from the inhabitants. Only with a healthy business sector that generates this income base will a municipality be able to collect enough resources to finance expanding social service programs for the quickly growing number of elderly.

Local Problems: The Administrator's View

Turning from political to administrative leadership, we present data obtained from the "problems" question in the questionnaire directed to local chief administrative officers. In accordance with the international

TABLE 3.6

Chief Administrators' Perceptions Regarding the Importance of
Factors Causing Financial Problems (in percentages)

Relevant Factors	Very Important	Somewhat Important	Least Important	Don't Know	N
State revenue loss	29	37	33	1	354
Inflation	15	52	32	1	355
Unemployment	17	34	48	1	358
Tax base decline	36	30	33	1	348
Service demands	21	48	28	3	357
State expenditure limits	4	14	79	3	354
Taxpayer pressure	3	16	80	1	356
Bond problems	6	10	83	1	357
State-mandated costs	35	48	16	1	359
Municipal employees	1	27	70	1	358
Lack of political priorities	23	38	37	2	353
Lack of administrative flexibility	4	26	67	3	353

SOURCE: Norway FAUI Project survey, 1985 chief administrator file.

FAUI questionnaire, this question was directly related to factors causing "financial problems" experienced by the municipality. Table 3.6
shows the assessments of how important various causes were perceived
to be based on answers from about 80% of the 454 CAOs in Norwegian
municipalities.

If we look solely at the "very important" column, we see that there
are two salient causes of financing problems perceived by the chief
administrators: tax base decline and state-mandated costs. If "somewhat important" responses are also included, however, state-mandated
costs are mentioned by a total of 83%, while tax base problems are
superseded by service demands (69% versus 65%). Other factors of
importance are loss of state revenue, inflation, unemployment, and lack
of political priorities. On the other side, few administrators give high
importance ratings to bond problems, state limits on income and spending, taxpayers' revolts, and pressures from municipal employees.

The administrators' views of the causes relating to financial problems confronted by Norwegian municipalities clearly identify the crucial role of state-local relations. As noted previously, the state has
transferred important welfare society responsibilities to the municipalities, and this has proven a costly experience for the local level.

As for other responses reported in Table 3.6, it is impossible to say whether problems related to the demands for services from local groups and citizens specifically refers to services for the elderly, the service area most frequently mentioned by the mayors. By attaching high importance ratings to tax base and unemployment considerations, however, the administrators' answers seem to coincide with and underscore the mayors' concerns about developing local industry and business opportunities.

That almost no significance is attached to municipal employees is not surprising, it might be added, inasmuch as general municipal salary schedules are negotiated at the national level. Taxpayers' resistance to spending and tax levels are also unlikely to be perceived as sources of major financial problems in a system where the state sets many service standards and standard tax percentage limits for all municipalities.

The last two items in Table 3.6 were specially added to the Norwegian version of the "financial problems" question. Administrators were asked to give their opinions as to whether lack of local political priorities and lack of administrative flexibility were factors relevant to the financial problems of their municipalities. A solid 61% responded that lack of local political priorities represented a consideration with important financial implications. Substantially fewer — a total of 30% — perceived organizational problems to be of importance for the municipal economy. These responses seem to suggest that local financial problems, at least as seen from the chief administrators' perspective, are more often of a political rather than administrative character.

Local Problems: Perceptions and Facts

The challenges reported by mayors, together with the sources of financial problems perceived by the chief administrative officers, give a fairly consistent picture of the Norwegian municipal scene in the mid-1980s. Local problems, as seen by the two types of leaders, are most frequently reported in the following areas:

(1) the nexus of business development-employment-tax base considerations, which involves voluntary municipal service provision as well as the administrator's more technically oriented function of monitoring personal income taxation

(2) political pressure for better service delivery in general, but especially concern for needs of the elderly

(3) state-local financial relationships (an area of concern primarily for the administrative leaders)

It is not possible to say to just what extent these perceptions are based on factual conditions. We know, for example, that 16% of the future tasks identified by the mayors concerned services for the elderly, while the corresponding figure for children's day-care centers was only 2%. We do not know, however, whether the mayors' strong emphasis on improving services for the elderly means that present performance in this area fails more dramatically than in the children's service sector.

Even though these perceptions may deviate from objective facts, it is nonetheless fair to presume that the way leaders act is in large part a function of their perceptions, not strictly a result of facts in any objective sense. With this in mind, we may now turn to the question of what the strategic choices are that have been made in trying to cope with the conditions experienced by Norwegian municipalities in recent years.

Strategic Choices in Norwegian Local Government

Municipalities can choose among a wide variety of strategies in attempting to deal with pressing concerns. The Norwegian questionnaire sent to chief administrative officers included 34 different strategies: 8 revenue-oriented strategies, 19 expenditure-oriented strategies, and 7 management-oriented strategies. Table 3.7 shows the most frequently used strategies reported by the CAOs. (Marginal frequency responses for all 34 strategies may be found in the Appendix to this chapter.)

As is evident from Table 3.7, the strategy that has most commonly been employed involves *increasing user fees*. It has been adopted by almost all municipalities (92%). Other revenue-oriented strategies that have been used frequently are lowering of surpluses, selling of some assets, and increasing long-term borrowing. All of these strategies have been adopted by roughly 60% of all municipalities.

The second most common coping strategy, however, is an expenditure-oriented strategy — *cutting expenditures in selected departments*. This strategy has been adopted by 81% of all municipalities. Three other expenditure-oriented strategies reported by at least half of all the chief administrative officers involved deferring maintenance of capital stock (66%), postponement of investments (51%), and reduction of expenditures for supplies, equipment, and travel (50%).

TABLE 3.7

Coping Strategies Adopted by Norwegian Municipalities (N = 372)

Strategy	%
Revenue-oriented strategies	
increase user fees	92
lower surpluses	61
sell assets	60
increase long-term borrowing	59
Expenditure-oriented strategies	
cut spending in some departments	81
defer maintenance of capital stock	66
postpone investments	51
reduce expenditures for supplies, equipment, travel	50
Productivity and management-oriented strategies	
introduce labor-saving techniques	61
improve productivity through better management	56

SOURCE: Norway FAUI Project survey, 1985 chief administrator file.

Among strategies focusing on improved productivity and management, two in particular have been adopted by a majority of Norwegian municipalities: improved productivity through introduction of labor-saving devices (61%) and improved productivity through better management (56%).

That such strategies are adopted, of course, may be of limited significance in terms of reducing perceived fiscal stress. Such strategies may be adopted for any number of reasons. For whatever reasons they are adopted, moreover, they may not be deemed very important in dealing with the problems at hand. For this reason, chief administrative officers were also asked to assess the relative importance of each coping strategy. Table 3.8 contains the distribution of responses for each of the most commonly employed coping strategies. As is evident from this table, a number of strategies are not judged to be very important in terms of reducing fiscal stress. Increased user fees, for example, although the most frequently used strategy in all municipalities, appears to be of limited importance for local governments as a means of reducing fiscal stress. In only 23% of the municipalities that have adopted this strategy did the CAO judge it to be a "very impor-

TABLE 3.8

Importance of Coping Strategies in Terms of Reducing Fiscal Stress in
Norwegian Municipalities (in percentages)

Strategy	Very Important	Somewhat Important	Least Important	Don't Know	N
Revenue-oriented strategies					
increase user fees	23	49	27	2	342
lower surpluses	20	51	25	4	228
sell assets	18	37	44	0	223
increase long-term borrowing	44	40	14	2	218
Expenditure-oriented strategies					
cut spending in some departments	55	38	5	2	301
defer maintenance	15	55	26	4	245
postpone investments	40	44	14	2	190
reduce expenditures for supplies, equipment, travel	11	66	21	2	186
Productivity and management-oriented strategies					
introduce labor-saving devices	36	52	11	1	225
better management	38	51	8	3	208

SOURCE: Norway FAUI Project survey, 1985 chief administrator file.

tant" strategy, whereas 49% indicated it was "somewhat important" and
fully 27% suggested it was "least important."

It can be noted here that this response pattern may reflect the fact that
user fees do not mean much when it comes to most local government
budgets (see Table 3.2, above, and Kleven & Rose, 1984). In 1985, for
instance, user fees accounted for no more than 9% of total revenues for
all Norwegian municipalities. Even so, it might nevertheless be argued
that such user fees may be of importance as a source for critical "top
financing" of services, and in maintaining service levels rather than
reducing them. To judge from the responses of chief administrative
officers, however, this is at best only marginally true.

The cutting of expenditures in selected departments, increases in
long-term borrowing, and postponement of investments, by contrast,
are all reported to have been quite important strategies in coping with
financial problems. Well over 80% of the chief administrative officers
in municipalities that have adopted these measures see these strategies
as having been very or at least somewhat important. Both of the

productivity and management-oriented strategies are also perceived as having been nearly equally important.

As noted previously, insofar as Norwegian municipalities have experienced fiscal crisis, it has generally been a crisis of relative rather than absolute character. The coping strategies most commonly employed seem to be in keeping with such a description. That is, most of the strategies adopted may be interpreted to suggest that municipalities have seen the fiscal squeeze as a temporary, short-term problem, not as a lasting one. Three of the most broadly used expenditure strategies (deferred maintenance of capital stock, postponement of investment, and reduction of expenditures for supplies, equipment, and travel), for example, are in essence *postponement* strategies, not genuine *reduction* strategies.

Likewise, all of the most commonly used revenue strategies except increased user fees are measures that mean a reduction of capital reserves and a growth in debt, not the creation of new revenue sources or expansion of existing ones. The municipalities, in short, appear to have emphasized strategies that will help them achieve a redistribution of resources between service sectors and secure some additional resources to meet demands during a short-term period of scarcity rather than adopting more hardhanded strategies involving service reductions.[7]

There is, of course, a great deal of variation in terms of both the strategies various municipalities have adopted and the number of strategies adopted. Different municipalities use different measures, or different combinations of measures. Only two strategies are used by more than 70% of all municipalities, while all strategy options mentioned in the questionnaire are used by at least some municipalities (see Appendix). Only four strategies have been used by less than 10% of the municipalities, three of these being measures that imply a reduction of the work force. Reduction of the municipal work force, however, is neither a popular nor a viable option for Norwegian local government. Municipal employees generally enjoy highly secure employment contracts and are organized in strong unions that help guarantee the enforcement of these contracts.

Most of the measures are used by between 20% and 60% of all municipalities. This situation suggests a great deal of variation in which strategies local governments have adopted. As Table 3.9 reveals, there is also a good deal of variation in the number of strategies each municipality has employed. The median municipality reports using 12 different strategies, and more than 60% of the municipalities have adopted between 7 and 15 strategies.

TABLE 3.9

Frequency Distribution for the Number of Strategies Adopted by
Norwegian Municipalities

Number of Strategies	%	N
0-3	5	17
4-6	11	40
7-9	20	73
10-12	25	92
13-15	17	64
16-18	11	41
19-25	6	24
26-34	6	21
Total	100	372

SOURCE: Norway FAUI Project survey, 1985 chief administrator file.

Explaining Strategic Choices

What factors can explain why some municipalities have adopted
many of these coping strategies, while others use fewer? Can differences in the number of strategies adopted be explained by specific
characteristics of particular municipalities?

In keeping with other work reported in this volume, we are in
particular interested in determining the relative importance of fiscal
stress experienced by municipalities compared to other factors that also
are of theoretical relevance in explaining these choices. In order to do
this we employ a causal model with seven independent variables — one
variable designed to measure fiscal stress plus six others.[8] For this
purpose fiscal stress is measured as the percentage change in gross
municipal tax revenues over the period from 1980 to 1983.[9] The six
other variables are as follows:

(1) size of the municipality, as measured by the number of inhabitants on
    January 1, 1983
(2) local wealth, as measured by average per capita taxable income in 1983
(3) local political ideology, as measured by the percentage of socialist party
    representatives on the local municipal council during the 1983-87 coun-
    cil period
(4) local mayoral spending preferences, as measured by the mayor's mean
    answer regarding personal preferences for spending more or less money
    on 20 different policy areas[10]

TABLE 3.10

Regression Analysis Results for Three Coping Strategy Indices (beta coefficients)

		Dependent Variables	
Independent Variables	*Revenue Index*	*Expenditure Index*	*Productivity/ Management Index*
Population size	−.056	.177	.024
Local wealth	−.036	−.291	.039
Local political ideology	.023	.031	.117
Mayoral spending preferences	.187*	.157	.011
Administrative modernity	−.011	.061	.183*
Central government transfers	−.155	−.136	.071
Fiscal stress	−.015	−.083	−.002
a (constant)	1.354	1.221	−.462
R^2	.042	.099	.049

*Significant at .01 level.

(5) local administrative modernity, as measured by the municipality's use of six fiscal management tools[11]

(6) level of central government transfers, as measured by the percentage of gross municipal income derived from central government grants in 1983

The dependent variables in this analysis, the strategic choices, are split into three summary indices representing the number of revenue-oriented, expenditure-oriented, and productivity/management-oriented strategies adopted by the municipalities.[12]

Results from the regression analyses using the same basic prediction model for each strategy index are presented in Table 3.10. Logarithmic transformations of all variables have been carried out in each case prior to running the analyses.[13] As we see from Table 3.10, our model does not provide a very good overall explanation of the number of strategies that have been adopted by any given municipality. Regardless of whether we consider revenue-oriented strategies, expenditure-oriented strategies, or productivity/management-oriented strategies, the model accounts for relatively little variation in the summary indices we have created for these alternative coping strategies. At best, the seven independent variables account for only 10% of the variation in the number of expenditure-oriented strategies adopted, whereas for the other two indices R^2 is less than 5%.

In keeping with this situation, the beta coefficients for any given independent variable are all of a very modest character. Only in two instances do they achieve a magnitude that, under the circumstances, would suggest significance at the .01 level.[14] These conditions make it difficult, if not impossible, to attribute much meaning to individual coefficients. Several aspects of the findings in Table 3.10 may nevertheless be highlighted for purposes of general interest with respect to subsequent work.

First, regarding the adoption of revenue-oriented strategies, the two independent variables of apparently greatest consequence are mayoral spending preferences and central government transfers. In each case the effect of these variables is much as might be expected: The more mayors express preferences for higher spending levels across a range of policy areas, the greater the number of revenue-oriented strategies municipalities are likely to have adopted (a response to mayoral preferences and leadership, perhaps?), and the lower the level of central government transfers to a municipality, the greater the number of revenue strategies the municipality is likely to have adopted (a "fill-the-gap" response, perhaps?). All the other variables appear to be virtually meaningless when seen in this context.

The same two variables are also of some consequence with respect to the adoption of expenditure-oriented strategies, although in a more attenuated sense. That mayoral spending preferences should be *positively* associated with the adoption of a greater number of expenditure-reduction strategies, moreover, is something of a paradox. One possible explanation might be that mayors with strong spending preferences are active in agitating for adoption of certain expenditure-reduction strategies in order to make more money available for other activities (among them, the mayor's pet spending projects), but this can only be a speculative line of conjecture.

The independent variable of greatest apparent consequence in this instance, however, is our measure of local wealth. Municipalities in which there is a stronger tax base, it would appear, have less need to adopt expenditure-reduction oriented strategies — a possibility that is quite plausible. Municipalities of a larger population size also show a slight propensity to adopt more expenditure-reduction strategies, although why this should be the case is less obvious. One possibility may be that it is in larger municipalities that tendencies toward bureaucratic expansion and excess are most likely to occur (see Downs, 1967; Niskanen, 1971), and hence the need for such expenditure-oriented strategies is greatest. But again, this is merely speculation at this point.

By extension it could be argued that this last line of reasoning, if valid, should also apply to the likelihood that municipalities would adopt productivity- and management-oriented strategies, that is, larger municipalities would have a greater need to adopt more such strategies in order to counteract bureaucratic excess. Yet the results in Table 3.10 do not bear out such an argument. Rather, the two independent variables of greatest apparent consequence for the adoption of such strategies are administrative modernity and local political ideology. The former is a consideration that, by virtue of its character, may quite reasonably be expected to be related to the adoption of productivity- and management-oriented strategies; they are, one might say, two sides of the same coin.

That local political ideology should be of modest consequence here, by contrast, is more surprising — in particular because of the direction of the apparent relationship. As we have measured this variable, the positive sign of the coefficient suggests that municipalities where the local councils have a higher proportion of socialist party representatives have shown a *greater* propensity to adopt productivity- and management-oriented strategies. This is contrary to conventional wisdom, which says that bourgeois or nonsocialist parties tend to be more oriented toward productivity and management concerns than their socialist counterparts. We have found evidence elsewhere, however, that suggests such common images are not always appropriate for Norwegian local politics (see Aarsæther, 1986b).

Perhaps the most noteworthy finding in Table 3.10 is nevertheless the virtually total lack of any relationship between our measure of fiscal stress and the three strategy indices when the other independent variables are also included in the prediction model. This finding may of course be due to the method by which we have operationalized this variable. We do not believe this to be the case, however. We are more inclined to believe that a similar finding would emerge using a wide range of measures for this variable. As we have tried to make clear above, after all, Norwegian municipalities have not been subject to fiscal stress in any objective sense. To the extent fiscal stress has been experienced, it has been of a marginal character; it has been moderate and short term in nature, not drastic or long term. Hence the absence of any significant effect of this variable on alternative coping strategies is quite understandable, and not surprising at all.

Affluence, Grant Structure, and Autonomy: Hypotheses for Further Research

Table 3.10 shows that not only fiscal stress but all the independent variables are weakly related to the choice of strategies for balancing municipal budgets. Why should these theoretically relevant variables be of such marginal influence in Norway?

We think there are two mechanisms at work that may account for this result. First, municipal affluence serves to "decouple" decision making from many external and internal constraints that would otherwise influence strategic choices. Choices may therefore be more a result of incidental and idiosyncratic factors pertaining to the participants and their organizational settings rather than the products of systematic contingencies that we have tried to map here. In a situation of affluence, furthermore, making strategy choices and thinking about balancing the budget may not be highly developed policy activities, but rather may form a residual field of policymaking. Decisions may be made more as the indirect product of other policies, not as the outcome of deliberate choice. If this is the case, metaphors such as "disjointed incrementalism" (Lindblom, 1965) and "garbage cans" (March & Olsen, 1976) may provide better models of the policy process than rational contingency thinking, which tends to emphasize policy constraints more than either of these alternatives.

Second, a categorical grant system may "distort" local decisions. It was shown above that national grants became an increasingly important source of local government financing during our period of study. These grants were earmarked for particular services and expressed the central government's ideas about what local policies should be, not necessarily the municipalities' own priorities. The function of categorical grants may therefore be not only to cushion municipalities against local constraints and demands, but also to standardize processes of decision making, thereby making them more immune to the influence of local factors.

Both of these hypotheses call for truly comparative research in order to be tested adequately. We would in particular expect the explanatory model used here to work better under conditions where there is (a) more overall fiscal stress, and (b) fewer categorical grants than has been the case for Norwegian municipalities during our period of study.

It is also appropriate to note in closing that recent changes in the Norwegian grant structure will probably affect municipal decision making and patterns of local influence in a significant fashion. We have referred previously to the abolition of the categorical grants system that took place in 1986. Since that time municipalities have received only one annual "paycheck" from the central government. No one would contest the idea that municipalities, as a result of this change, enjoy more discretion in the use of available economic resources. Preliminary observations from case studies, however, indicate that the new block grant system has so far resulted in more influence for the chief administrative officer, no change in influence for the political parties, but less influence for both the municipal council and local interest groups (Aarsæther, 1986a).

The transition from a system of categorical grants to a block grant system has also been a painful experience for several municipalities, especially where leaders had acquired a high degree of competence in dealing with categorical grants (see Aarsæther, 1987; Sørensen, 1987). More local autonomy has been achieved under the new grant system, in short, but one may ask whether the price will prove to be too high given possible increases in administrative as opposed to political influence, and additional pressures placed on local processes for setting fiscal priorities.

APPENDIX

Frequencies for 34 FAUI Strategies in Norwegian Municipalities

Strategy	%	N
Revenue-oriented strategies		
seek new local revenues	36	135
seek additional intergovernmental revenues	40	149
increase taxes[a]	33	121
increase user fees	92	342
lower surpluses	61	228
sell some assets	60	223
increase short-term borrowing	24	90
increase long-term borrowing	59	218
Expenditure-oriented strategies		
defer payments to next year	29	108
cut spending for all departments	22	81
cut spending for least efficient department[b]	81	301
lay off personnel (temporarily)	5	20
dismiss personnel[c]	8	31
reduce administrative expenditures	26	98
reduce employee compensation[d]	27	101
freeze hiring	33	123
reduce work force through attrition	26	97
reduce expenditures for supplies, equipment, travel[e]	50	186
reduce services funded by own revenues	23	85
reduce services funded by intergovernmental grants	16	61
eliminate programs	20	73
reduce capital expenditures	35	130
keep expenditure increases to inflation	35	130
early retirements	7	25
reduce overtime	27	101
defer maintenance of capital stock	66	245
postpone investments[f]	51	190
freeze wages and salaries[g]	—	—
control construction to limit population growth[g]	—	—

APPENDIX
(continued)

Strategy	%	N
Productivity and management-oriented strategies		
shift responsibility to other units of government[h]	22	83
contract out services to other units of government[i]	39	145
contract out services to private sector[j]	17	65
involve voluntary organizations in problem solving[k]	39	144
improve productivity through better management	56	208
improve productivity through labor-saving techniques	60	225
make purchasing agreements[l]	8	28

a. In Norway it is only possible for the municipalities to increase property taxes. The item was therefore phrased "Increase property taxes" on the Norwegian questionnaire (modification of standard FAUI item).

b. The item was phrased "Budget cuts for some departments" on the Norwegian questionnaire (modification of standard FAUI item).

c. This is a country-specific item added to the Norwegian questionnaire in order to sharpen the distinction between permanent and temporary employee-reduction measures.

d. The item was phrased "Reduce employee participation in seminars, conferences and the like" on the Norwegian questionnaire (modification of standard FAUI item).

e. The item was phrased "Reduce purchase of equipment and supplies" on the Norwegian questionnaire (modification of standard FAUI item).

f. This is a country-specific item added to the Norwegian questionnaire.

g. These two items were excluded from the Norwegian questionnaire.

h. The item was phrased "Shift responsibility for specific tasks to state- or county-level agencies" on the Norwegian questionnaire (modification of standard FAUI item).

i. The item was phrased "Get the state/county to cooperate in solving specific tasks" on the Norwegian questionnaire (modification of standard FAUI item).

j. The item was phrased "Greater use of private firms" on the Norwegian questionnaire (modification of standard FAUI item).

k. This is a country-specific item added to the Norwegian questionnaire.

l. The item was phrased "Enter into intermunicipal purchasing agreements" on the Norwegian questionnaire (modification of standard FAUI item).

Notes

1. Under the circumstances, municipalities of the former variety, that is, those with high service aspirations, were clearly more to the liking of ministries such as health and education, for example, whereas municipalities of the latter variety were much preferred in the eyes of the Ministry of Finance (see Aarsæther, Larsen, & Rovik, 1981).

2. These goals were explicitly identified by a commission on local government reform appointed by the central government (see Norges offentlige utredninger, 1974).

3. Data from our surveys of mayors and chief executive officers showed an overwhelmingly positive attitude toward this new grant system, at least prior to its introduction. Some recent findings, however, suggest that not everyone is equally enthusiastic at this point (see, e.g., Aarsæther, 1987; Sørensen, 1987).

4. The central government is capable of influencing municipal borrowing behavior in two primary fashions. One is by means of adjusting interest rates and making decisions on loans available to the municipalities through a state-run Municipal Bank. The other is a more conditional control via the county prefect's budgetary review procedures. All municipalities have traditionally been required to submit their annual budgets to the prefect for review and approval, a routine designed to assure that the municipalities establish budgets that, in theory at least, are based on reasonable economic suppositions and balanced as required by law. To the extent that a municipality is to receive a tax equalization grant, the prefect is empowered to refuse to approve the budget if it is balanced by means of proposed borrowing. Central authorities may also deny approval for borrowing from nondomestic sources, but otherwise have no direct means of preventing municipal borrowing from domestic sources should municipalities be willing to pay the price.

5. The major responsibilities currently assigned to municipalities by law have to do with primary education, primary health and social services, preschool institutions (nurseries and kindergartens), certain cultural activities, land use planning, certain control functions (e.g., fire protection, pollution and sanitation control services), and general municipal administration. County governments, by contrast, have principal responsibility for secondary education, the operation of hospitals and other major medical institutions, and various communication services (especially highway development and maintenance, plus the operation of bus and ferry services). Aside from these stipulated responsibilities, Norwegian law also permits units of local government to undertake other functions and activities that are not expressly forbidden or explicitly vested in other institutions of government (a negative delimitation of authority, as noted previously).

6. The fact that mayors also mention communications as a relevant task may be understood in terms of the country's geographical structure and an incessant public discussion about how different areas of the country can be provided relevant services on an equal basis.

7. To the extent that the fiscal squeeze is of short duration, a temporary aberration rather than a permanent state of affairs, these strategic choices might well be seen as a form of rational behavior. This is a line of argument that we are not able to pursue here, however.

8. The model employed here is adapted to specific conditions pertaining in Norway. Thus some measures that have been included in earlier work on questions of this sort (e.g., Clark, Burg, & de Landa, 1984) or elsewhere in this volume — measures relating to the ethnic composition of the population, the percentage of the population that is foreign born or the percentage living below the poverty level, and local political cultures, for example — are not included in our model. The variables that are included are nonetheless chosen and operationalized in a fashion intended to facilitate rudimentary cross-national comparisons.

9. The concept of fiscal stress has been subject to much discussion (see, for example, Clark & Ferguson, 1983; Eberts, 1985; Mouritzen, 1985, 1987). The measure employed here is limited solely to fiscal considerations — that is, political demands and resources, or other nonfiscal aspects, are not incorporated in our measure — and the measure is fashioned to capture relative shifts in the local resource base as represented by changes

in *locally derived* income tax revenues. Such local tax revenues are, as noted previously, the primary source of revenues available to municipalities for discretionary spending. Intergovernmental transfers, the other major source of revenues, were, until recently at least, increasingly of a categorical, earmarked variety that gave municipalities less autonomy in terms of their spending decisions. Hence we would argue that a decline in locally derived tax revenues during the period from 1980 to 1983 is likely to have been perceived as a threat to local fiscal autonomy and therefore a potentially reasonable measure of fiscal strain, to the extent this was experienced at all.

10. To receive a score on this index, respondents had to have valid answers for at least 10 of the 20 policy areas.

11. This index is based on FAUI questionnaire items regarding six dimensions of fiscal management (revenue forecasting, fiscal monitoring, performance measurement, accounting and fiscal reporting practices, local economic analysis, and fiscal reporting to the municipal council), each of which permits varying degrees of administrative "modernity" or sophistication. The index consists of a simple mean of the chief administrative officer's responses to individual items, with valid responses on a minimum of four out of the six items required in order for the municipality to receive an index score. For additional details regarding construction of this index, see Rose (1986, p. 9).

12. The strategy indices employed here are computed as the simple sum of relevant strategies used in the municipality. In the event a respondent left certain strategies blank while responding positively to others, these blanks were given a value of zero. A presumption was made, in other words, that strategies left unmarked were not used. Cases in which respondents have not indicated the use of *any* strategies and left the entire battery blank, however, have been treated as missing data and have been omitted from the analysis.

13. Rather than using a simple additive model, we adopted this analytic approach as a means of correcting for the somewhat skewed distributions of the dependent variables. Results obtained for analyses based on a simple additive model, however, are essentially the same.

14. It is perhaps relevant to stress here that inasmuch as our data are based on a universe of cases, not a sample (i.e., all municipalities were surveyed, and over 80% replied), all empirical results may be argued to be of substantive significance in the sense that they are not subject to problems of sampling error. The results may nonetheless be subject to random measurement error due to other sources, so significance tests retain some relevance in this respect.

References

Aarsæther, N. (1985). Kommunalt tiltaksarbeid — eit virkemiddel med vanskelege oppvekstvilkår. In H. O. Larsen & N. Årsæther (Eds.), *Kommunalt tiltaksarbeid.* Tromsø: University Press.

Aarsæther, N. (1986a). *Kommunalpolitisk budsjettering.* Unpublished doctoral thesis, University of Tromsø, Institute for Social Sciences.

Aarsæther, N. (1986b, April 1-6). *Political learning and administrative rationality: The transformation of classical social democracy.* Paper prepared for the Political Learning, Fiscal Austerity and Urban Innovation workshop, ECPR Joint Sessions of Workshops, Gothenburg, Sweden.

Aarsæther, N. (1987). *Kommunal beslutningsstruktur og nytt inntektssystem* (Working Paper No. 3, New Transfers System Project). Tromsø: University of Tromsø, Institute for Social Sciences.

Aarsæther, N., Larsen, T., & Rovik, K. A. (1981). *Velferdsstaten som kommunaløkonomisk problem* (Report No. 2, Municipal Government Problems Project). Tromsø: University of Tromsø, Institute for Social Sciences.

Baldersheim, H. (1986, April 1-6). *The choice of fiscal management strategies: The role of situational, organizational and personal factors for innovation.* Paper prepared for the Political Learning, Fiscal Austerity and Urban Innovation workshop, ECPR Joint Sessions of Workshops, Gothenburg, Sweden.

Baldersheim, H. (1987). Frå statstenar til stifinnar: 1970-1987. In H. E. Naess et al., *Folkestyre i by og bygd: Norske kommuner gjennom 150 år.* Oslo: University Press.

Clark, T. N., Burg, M. M., & de Landa, M.D.V. (1984, August 30-September 2). *Urban political culture and fiscal austerity strategies.* Paper presented at the annual meeting of the American Political Science Association, Washington, DC.

Clark, T. N., & Ferguson, L. C. (1983). *City money: Political processes, fiscal strain, and retrenchment.* New York: Columbia University Press.

Downs, A. (1967). *Inside bureaucracy.* Boston: Little, Brown.

Eberts, P. (1985). Fiscal austerity and its consequences in local governments. In T. N. Clark (Ed.), *Research in urban policy* (Vol. 1). Greenwich, CT: JAI.

Fevolden, T., & Sørensen, R. (1987). Norway. In E. C. Page & M. G. Goldsmith (Eds.), *Central and local government relations: A comparative analysis of West European unitary states.* London: Sage.

Hansen, T. (1986a). *Fra vekst til innstramming: Finansiell utvikling i norske kommuner 1972-1982* (Report No. 3/86). Oslo: Rådet for forskning for samfunnsplanlegging (NAVF/RFSP), Kommunaløkonomisk forskning, Nytt inntektssystem for kommunene.

Hansen, T. (1986b). Kommunaløkonomiske problem og det nye inntektssystemet. In J. Naustdalslid (Ed.), *Kommunal styring: Innføring i kommunalkunnskap frå ein planleggingssynsstad.* Oslo: Det Norske Samlaget.

Hansen, T. (1987). *Financial development in Norwegian local government.* Oslo: University of Oslo, Institute of Political Science.

Kjellberg, F. (1981a). The expansion and standardization of local finance in Norway. In L. J. Sharpe (Ed.), *The local fiscal crisis in Western Europe.* London: Sage.

Kjellberg, F. (1981b, November 5-6). *Local government and the Scandinavian welfare state: Structural and functional reforms in Denmark, Norway and Sweden since 1945.* Paper presented at the Colloquy on the Reforms of Local and Regional Authorities in Europe: Theory, Practice and Critical Appraisal, Council of Europe, Linz, Germany.

Kjellberg, F. (1985). Local government reorganization and the development of the welfare state. *Journal of Public Policy, 5*, 215-239.

Kleven, T., & Rose, L. (1984). *The provision of urban services: A national overview* (NIBR Note 1984: 152). Oslo: Norwegian Institute for Urban and Regional Research.

Kommunal- og arbeidsdepartementet (1986). *Forsøksvirksomhet i kommunene og fylkeskommunene-frikommuner* (Rundskriv H-14/86). Oslo: Ministry of Municipal and Labor Affairs.

Larsen, H. O. (1985, March 25-30). *Mayoral leadership and problems of urban governance.* Paper prepared for the Metropolitan Government workshop, ECPR Joint Sessions of Workshops, Barcelona.

Larsen, H. O. (1987, April 10-15). *Political authority and local leadership.* Paper prepared for the Leadership and Local Politics workshop, ECPR Joint Sessions of Workshops, Amsterdam.

Lindblom, C. (1965). *The intelligence of democracy.* New York: Free Press.

March, J. G., & Olsen, J. P. (1976). *Ambiguity and choice in organizations.* Bergen: Universitetsforlaget.

Mouritzen, P. E. (1985). Concepts and consequences of fiscal strain. In T. N. Clark, G.-M. Hellstern, & G. Martinotti (Eds.), *Urban innovations as response to urban fiscal strain.* Berlin: Verlag Europäische Perspektiven.

Mouritzen, P. E. (1987). Dimensions of fiscal strain. In T. N. Clark (Ed.), *Research in urban policy* (Vol. 3). Greenwich, CT: JAI.

Niskanen, W. A. (1971). *Bureaucracy and representative government.* Chicago: Aldine-Atherton.

Norges offentlige utredninger. (1974). *Mål og retningslinjer for reformer i lokal forvaltning* (NOU 1974: 53). Oslo: University Press.

Norges offisielle statistikk. (1977). *Kommuneregnskap 1975* [Municipal accounts 1975] (Statistisk ukehefte No. 10/77). Oslo: Central Bureau of Statistics.

Norges offisielle statistikk. (1982). *Kommunenes regnskaper for 1980* (Statistisk ukehefte No. 10/82) [Municipal accounts 1980]. Oslo: Central Bureau of Statistics.

Norges offisielle statistikk. (1986a). *Kommuneregnskap 1985* (Statistisk ukehefte No. 50/86) [Municipal accounts 1985]. Oslo: Central Bureau of Statistics.

Norges offisielle statistikk. (1986b). *Nasjonalregnskap 1975-1985* [National accounts 1975-1985]. Oslo: Central Bureau of Statistics.

Norges offisielle statistikk. (1987). *Nasjonalregnskap 1976-1986* [National accounts 1976-1986]. Oslo: Central Bureau of Statistics.

Ot. prp. nr. 48 (1984-85). *Om endringer i lover vedrørende inntektssystemet for kommunene og fylkeskommunene.* Oslo: University Press.

Ot. prp. nr. 36 (1985-86). *Om midlertidig lov om utvidet forsøksvirksomhet i kommuner og fylkeskommuner (frikommuner).* Oslo: University Press.

Rose, L. E. (1986, April 1-6). *The use of fiscal management tools in Norwegian municipalities: A preliminary report.* Paper prepared for the Political Learning, Fiscal Austerity and Urban Innovation workshop, ECPR Joint Sessions of Workshops, Gothenburg, Sweden.

St. meld. nr. 26 (1983-84). *Om endringer i lover vedrørende inntektssystemet for kommunene og fylkeskommunene.* Oslo: University Press.

Strand, T. (1985). *Utkant og sentrum i det norske styringsverket.* Bergen: University Press.

Sørensen, R. J. (1987, November 25-27). *Ønsker kommunene større lokalt selvstyre?* Paper presented at the NORAS Seminar on Municipal Economy, Sundvolden, Norway.

4

Fiscal Policymaking in Times of Resource Scarcity: The Danish Case

POUL ERIK MOURITZEN

At the end of the 1970s it made sense to refer to "the affluent Danish local authorities" (Bruun & Skovsgaard, 1981). This characterization is no longer valid. To elected as well as appointed officials in Danish local government, the first part of the 1980s will always be remembered as the "lean" years when it became increasingly difficult to balance the budget.

The Extent of Fiscal Stress

Fiscal stress may be defined as "a situation in which a local government, faced with the necessity of achieving a balance between revenues and expenditures, must in time choose either (1) increase taxes through changes in the tax rate or structure in order to maintain real expenditure and service levels, (2) reduce real expenditures from the level of the previous year or (3) engage in some combination of these activities" (Wolman & Davis, 1980, p. 1; see also Mouritzen, 1987a). Taking this definition as a starting point, the extent of fiscal stress in Danish local governments is indicated in Table 4.1.[1]

The figures show how much expenditures (in constant prices) would have to change from the previous year in order to balance revenues and expenditures if the tax rates and fees and user charges (in constant prices) were left unchanged. A negative value is indicative of fiscal stress, because expenditures have to be cut and/or taxes have to be raised. A positive value signifies a situation where expenditures in real terms may be increased without increasing the rate of taxation. To take

TABLE 4.1

Fiscal Stress in Danish Local Government, 1978-1986

	1978	1979	1980	1981	1982	1983	1984	1985	1986
Change in expenditure with constant rate of taxation and fees	—		−3.9	−2.3	−0.9	−1.4	−2.6	4.5	−0.8	−5.3

NOTE: All figures are per capita, constant prices. They show percentage change in total expenditures necessary to balance revenues and expenditures if the rate of taxation, user fees and charges (in constant prices), liquid assets, and debt are left unchanged.

101

an example, the value of −2.6% for 1983 shows that if local governments had left tax rates and fees and charges unchanged from 1982 they would have had to cut expenditures by 2.6% to balance budgets.[2]

The fiscal stress time series shows that fiscal stress as defined above was a fact that local leaders had to live with in most of the period under investigation. In only one year — 1984 — were Danish local governments able to increase expenditures without raising revenues. In the remaining years, to maintain current levels of expenditures local officials had to raise taxes and fees and charges. However, the magnitude of these problems as indicated by the figures in Table 4.1 is not indicative of a major fiscal crisis. In only two years — 1979 and 1986 — the problem of balancing the budget required a reduction in expenditures of more than 3% or an equivalent increase in revenues. A calculation with 1978 as the base year shows that it would have been necessary to cut expenditures by close to 9% from 1978 to 1986, assuming fixed tax rates and fees and charges.

The Causes of Fiscal Stress

There are two major causes of fiscal stress in Danish local government: trends in the national economy and central government policies toward local governments. The importance of these two factors varies over the period.

Up until 1982-1983, local fiscal problems had their root in a national recession. This caused slow or negative growth in the local tax base and, at the same time, increased social welfare expenditures.[3] In this period central government, led by the Social Democrats, implemented a relatively "soft" policy of recommended, or nonmandatory, ceilings on growth in local expenditures. The aim of central policies in this period was mainly to keep down local taxes.

In the fall of 1982 a new conservative coalition government took over from the Social Democrats. The new government drastically changed policies toward local governments. In order to reduce the deficit in the national budget (in 1982 close to 10% of GDP, or 14% of total public expenditures), it implemented a series of cuts in the general grants to local governments. When the national economy entered a boom period in 1983, the conservative government introduced what was to become known as the "take-home principle." The argument was that the boom in the private sector was due to the economic policies of the government. Consequently, the government had the right to "take home" the resources that would otherwise flow into the local government sector. The way to do it was to reduce the grants.

TABLE 4.2

Tax Base and General Grants in Danish Local Government, 1978-1986
(index values, 1978 = 100)

	1978	*1979*	*1980*	*1981*	*1982*	*1983*	*1984*	*1985*	*1986*
Local tax base	100.0	101.3	102.6	101.9	103.2	104.6	105.5	107.3	110.8
General grants to local governments	100.0	93.0	98.6	103.7	107.8	96.0	94.5	57.7	32.8

NOTE: All figures per capita, constant prices.

These trends are depicted in Table 4.2, which shows that the local tax base increased slowly from 1978 to 1984, with a slight downturn in 1981. However, from 1984 to 1986 it rose sharply, about 5% in real terms.[4] The general grants, in contrast, were on the increase from 1979 to 1982, when they accounted for 13% of local revenues (see below). Since 1982 they have been reduced by 70% in real terms and now account for a mere 3% of local revenues.

Great Variations

An analysis of aggregate trends obviously conceals great variation among the 273 Danish municipalities. Some of them have never experienced fiscal problems at all. Others have been forced to implement large tax increases and/or cutbacks. The variation is shown for the period 1982-1986 in Table 4.3. The table shows that less than 4% of all local governments were not hit by fiscal stress as defined above; that is, they could raise operating expenditures without increasing tax rates or fees and user charges.[5] The remaining were hit to a varying degree, with most of the municipalities lying in the range between 5% and 15%. At the bottom we find approximately 17% of the municipalities that had to cut expenses by more than 15% over the four-year period or had to find equivalent revenues.

How did local governments react to fiscal stress? This is the major question addressed below. The chapter proceeds in three steps. To assist the reader in understanding the behavior of local political leaders, the next section is focused on the political environment of those leaders, with a special emphasis on relations to central authorities and local group processes. In the second section a closer look will be taken at the fiscal policies that have guided Danish local governments through these years of fiscal stress. Part of this analysis is based on aggregate

TABLE 4.3

Fiscal Stress Among Danish Local Governments, 1982-1986 (N = 273)

Degree of Fiscal Stress	Relative Frequency
< −20	7.0
−20 to −15	10.2
−15 to −10	24.2
−10 to −5	30.0
−5 to 0	17.2
0 to 5	7.7
> 5	3.7
Total	100.0

data, but some of it is based on city-level data. Finally, the consequences of this fiscal behavior of Danish local government for the distribution of power and values in society is discussed.

The Political Environment of Fiscal Policymaking

The three major elements in the political environment of local political leaders are the central government, the political parties, and local interest groups. Each is discussed in turn below.

The Central Government

The way the central government organizes the system of local government and local finances has very important implications for the local fiscal policymaking process. Also, the way central authorities control, regulate, and direct fiscal policymaking limits the strategies that are available to political leaders in times of fiscal stress.

Up until 1970, a system of detailed central government control and narrowly defined categorical grants persisted in Denmark, leaving little room for local autonomy. After more than ten years of planning, local government reform was implemented in 1970, starting with the amalgamation reform that reduced the number of municipalities from more than 1,200 to 275.[6] During the 1970s, decentralization of national programs to local governments also began. In the same period the system of intergovernmental grants was gradually changed. Many of the categorical grants were abolished and a system of general grants with a high degree of equalization built into it was introduced. This reform was perceived as an effort to give local officials more autonomy in the priority-setting process, as grants given without strings could be used for any purpose, even reduction of local rates of taxation.

The extent to which these decentralization efforts have been successful can be debated. On the one hand, the reforms have created a relatively simple and highly consolidated local governmental system. It is *territorially consolidated* because jurisdictional boundaries correspond very much to the sociological realities of city life ("one city, one municipality").[7] It is consolidated in *programmatic terms* because most public services at the local level are the responsibility of one authority, the municipality. Finally, the system is *functionally consolidated* because policymaking, financing, and service delivery are the responsibility of the same unit, at the local level the municipality.

On the other hand, local government policies are still controlled very much by central authorities. This control takes many forms, ranging from very detailed and binding rules to broad, noncommittal goals and recommendations. For the purposes of this study, three aspects of the local policymaking process in which central control may be exercised are important: programs, personnel, and finances.

Programs. The main functions of municipal government in Denmark fall into six categories, of which social and health services account for almost 50% of total expenditures, education and culture for 20%, and administration and planning for 10%. The remaining 20% are divided among urban development, public utilities, roads, and capital expenses. In Table 4.4, expenditures are disaggregated further into twelve functions. The table gives information about the functions' share of total local expenditures (importance), the share of expenditures within each function paid for by the local government (the rest being paid for by central government via categorical, matching grants) (fiscal responsibility), and, finally, the proportion of expenses for each function that—in the perception of local administrators—is bound by central government regulations (constraints). The twelve functions are ranked according to the last measure, the degree of constraint.

The two major tasks of local government are primary schools and programs for the elderly, which account for approximately one-third of total expenses. In comparison with most other countries, Danish local governments have very few capital expenditures—they account for a mere 6.2% of total expenditures. When it comes to fiscal responsibility,

TABLE 4.4

Importance, Fiscal Responsibility and Central Regulation of Twelve Programs

	Importance (% of total local expenditure)	Responsibility (% financed by local revenues)	Constraints (% bound by central regulation)
Social assistance	5.1	50	80.7
Primary schools	15.5	100	78.4
Unemployment programs	4.6	–	74.5
Administration	9.9	100	64.7
Collective transportation	1.0	100	60.8
Social counseling	–	100	59.0
Day-care institutions	8.4	62	58.3
Programs for elderly	14.7	41	57.1
Public utilities	5.9	100	57.1
Libraries	1.5	100	55.3
Sport, culture	3.8	100	34.5
Roads	2.9	100	33.2
Other operating expenditures	20.5	–	
Capital expenditures	6.2	100	
Total	100.0	77	

SOURCE: Kommunalstatistiske meddelelser, 09-11 and 09-12 (1985). Based on 1985 budgets, excluding pensions.
NOTE: Last column shows the percentage of total costs of the various programs that are, in the perception of the city directors (N = 39), bound by central government regulations.

the general rule is that most programs are financed 100% by local revenues or by general block grants (including also capital expenditures). Within the social and health services there are still a number of matching grant schemes in existence, as indicated, for instance, by the 50% central government fiscal responsibility for social assistance. The final column in Table 4.4 reveals the perceptions of the city directors. Although these figures cannot be taken as objective evidence about the degree of central government control through regulations, the ranking does seem to make intuitive sense.[8] At the top of the list are three programs — social assistance, primary education, and unemployment programs — where there seems to be relatively tight central control of local activities. At the bottom of the list are some programs for which local autonomy is very wide: roads, sports, and cultural activities.

Table 4.4 further indicates that fiscal responsibility and regulatory control are not necessarily correlated. In terms of fiscal responsibility as well as regulatory control, social assistance is truly a "low-autonomy" program. But programs for elderly and day care must be

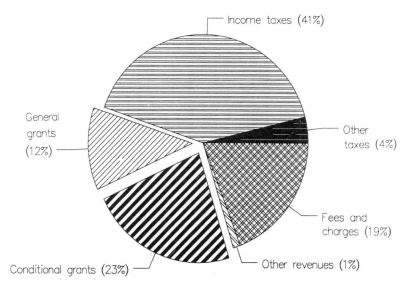

Figure 4.1 Local Revenues by Source, 1982 (final accounts, percentages)

categorized as "low-autonomy" when it comes to fiscal responsibility, while they are "moderate-autonomy" programs in terms of regulatory control.

Personnel. The Danish public personnel system is highly centralized. Salaries and general working conditions are negotiated centrally between unions and the associations of local authorities, but in reality decisions on the employers' side are made by the minister of finance. Pension schemes also are part of these negotiations. This system of uniform wages, working conditions, and pension arrangements is designed mainly to limit competition among public authorities for qualified personnel. Obviously, it also limits the strategies available to local governments during times of fiscal stress. What remains, of course, is the fact that the *size* of the municipal work force is still a local decision.

Finances. By far the most important source of revenue for Danish municipalities is the local income tax. In 1982 this accounted for 40% of total revenues (Figure 4.1). The system is based on proportional taxation, and taxes are levied on the same base as the county and central government income tax. *Local authorities are free to set their own rate of taxation.* The local rate varies between 13% and 22%, with an average of about 18.5%. For most income categories the sum of total regional and central rates varies between 50% and 60%.

Income taxes are paid by the source and are collected as an aggregate sum on the monthly paycheck. The wage earner is therefore not made aware, on a regular basis, of the proportion going to the separate jurisdictions.

Grants from central government (excluding pension reimbursements) accounted for 36% of total revenues in 1982. About two-thirds of these were in the form of conditional matching grants — exclusively within the social and health services. The importance of conditional grants is declining: They accounted for 13% of total revenues, but since 1982 they have been reduced substantially (Table 4.2).[9]

The Political Parties

The idea of local autonomy is linked to broader societal issues of freedom and equality. Such normative issues form one of the main bases of differentiation between the political parties. Survey findings indicate that the attitude toward local self-determination closely follows the classical left-right dimension in the Danish party system (Table 4.5). Conservatives tend to prefer local autonomy to equalization, socialists exhibit the opposite position, and the Social Democrats are in between. It follows that the party system and the political parties themselves potentially are very important political structures for the understanding of central-local relations.

Local elections are held every fourth year based on proportional representation; that is, each party or nonpartisan list obtains seats in the city council in proportion to the votes carried. In addition, representation on the important local finance committee and various standing committees is based on proportional votes in the council. Each municipality is a single district in electoral terms. While nonpartisan lists used to play an important role in small rural municipalities, the national political parties have gradually gained a near monopoly via their district and municipal organizations. Today only 15% of all elected officials in local government are not elected on a partisan list, but they are generally identifiable in terms of position on the traditional left-right continuum.

Consequently, there are few chances for an independent political entrepreneur to be elected. Nomination from within the party is normally prerequisite to a seat in the council. Nomination comes from party recognition, which is usually the result of years of loyal party activity. The individual owes more to the party organization and its platform

TABLE 4.5
Local Autonomy Versus Equality and Party Affiliation (in percentages)

	Emphasize Equality	Emphasize Autonomy	Neutral	N
Socialists	56.9	33.2	9.9	147
Social Democrats	40.8	43.0	16.2	686
Conservatives	27.7	57.0	15.3	705

SOURCE: Bentzon (1981, p. 310). Based on representative sample of Danish electorate. Socialists are respondents who voted for the Communists, Leftist Socialists, or People's Socialist party. Conservatives include people who voted for the Radical Liberals, the Conservative party, the Centre Democrats, Christian People's party, the Liberals, and the Progress party.
NOTE: Respondents were classified as emphasizing equality if they agreed with the item: "Central government should see to it that differences in services and taxation do not become larger than today." Respondents for autonomy were those who agreed with the item: "The local voters should decide how much service they want and what they want to pay in local taxes." Respondents had to choose between the two items.

than to his or her particular stand on issues. Often the candidate does not depend on the local party for success at the polls; election results are more likely to be a function of the success of the party at the national level. A landslide in a national election may have important consequences in the next local election.

Local political parties are parties in their own right, but they are also segments of national parties. The large national parties have an organization that closely mirrors that of national, regional, and local governments, with district, municipal, county, and national segments. Local parties may be perceived as representatives of nationwide parties with a view to diffusing viewpoints at the local level. The central parts of the party as well as individuals and groups of the party, who possess positions of power at the central level, function as channels through which the local parties may seek influence.

Two parties dominate local politics—the Social Democratic party and the Liberal (or Agrarian) party. Of the total mayoral positions, 77% are held by these two parties—27% by the Social Democrats and 50% by the Liberals.

The major lines of cleavages are clearly between the political parties. The city directors of 39 municipalities were asked in the 1982 survey about the major cleavage lines. While 80% found the partisan dimension to be important, only 23% reported important splits based on territorial affiliation of councillors (Dilling Hansen, 1983, p. 203).

The partisan differences concerning citizens' preferences for local autonomy (Table 4.5) are reflected also when it comes to the coherence of the political parties. The Social Democratic party acts more as a "national" party, while in the conservative parties – including the Liberal party – local branches perhaps are more independent. For instance, the importance of the party program for the behavior of the local party differs. About 70% of the Social Democrats in a 1983 survey of the local party branches reported that they tried to act according to the party program. Only about 30% of the local branches of the Liberal party reported similar behavior (Villadsen, 1985a, p. 174). Also, the Social Democrats are more likely to lobby national (party) leaders than are conservative local politicians.[10] Finally, party discipline within the local groups is more pronounced among the leftist parties than among the bourgeois parties (Mouritzen, 1985a, p. 27).

The Constellation of Local Interest Groups

The fact that local political leaders are deeply tied to the national party organization may have important consequences for their choice of strategies during times of fiscal stress. But they are also responsive to a local political environment consisting of individuals and groups not necessarily linked to the parties. In Denmark this local political environment contributes to a fiscal policymaking process biased in favor of greater public expenditures or minimally against major expenditure cuts. This bias is rooted in several characteristics.

The high degree of consolidation discussed above has several implications. In an average Danish municipality more than 50% of the electorate will be highly dependent on municipal services. An average of 20% of the electorate is directly dependent in that either husband or wife is a municipal employee. Another 20% of the electorate is made up of elderly people, who may receive help through old-age pensions, visiting health nurses, home help, or house rent support, or who may live in municipal nursing or sheltered homes. Up to 10% of the electorate is periodically or permanently highly dependent – requiring social assistance or sick allowances – or participating in municipal employment programs. Families with children are also dependent in their daily lives on municipal day-care institutions (more than 50% of youngsters up to 6 years old attend municipal day-care programs). The fact that a large proportion of the electorate is highly dependent on municipal services is likely to reduce the incentive of political leaders to implement major reductions in services. Also, the high degree of territorial and programmatic consolidation represents a disincentive to citizens and employees to oppose tax increases. In a highly consolidated system, the relationship between levels of taxation and spending for an individual program is obscure and uncertain. If one is dependent on municipal services, it may be dangerous to oppose tax increases because resource scarcity may threaten those services. This phenomenon is further strengthened by the relative invisibility of taxes. Although income taxes are generally conceived of as a relatively visible revenue source, the amount of money paid to a Danish municipality is obscured by the "withholding" provision and the system of tandem collection of state, county, and municipal taxes.

Finally, the high degree of consolidation (territorial, programmatic, and functional) tends to create interest groups that protect services. Several factors favor the mobilization of citizens and employees when particular programs are threatened: a high degree of professionalism, a high level of organizational membership in unions, and the existence of user councils within various service delivery institutions. There tends to be a strong congruence between the interests of providers and consumers in building a forceful constituency for maintaining expenditure levels. This congruence is strengthened further by the centralized system of wage negotiations, which has in effect removed one potential source of conflict, the higher wages versus more jobs issue. The high degree of territorial and programmatic consolidation results in a union structure where a number of relatively strong unions operate within the same territorial boundaries. Although they are affiliated with different programs, they have a strong incentive to form coalitions that cut across the various areas.

While there are strong forces in the Danish system that encourage the creation of interest groups around expenditure issues, there are very few organized channels of influence for citizens who are dissatisfied with taxes. At the local level, revenue decisions (taxing, borrowing, and the use of liquid assets) are made solely by the political and administrative elite in a relatively closed system of decision making. There are no provisions for binding referenda upon fiscal decisions, nor any provisions for citizens' initiatives. This particular feature of the decision-making structure makes it very difficult for the individual opposing high taxes to voice a protest.

Since the initiation of the local government reforms in 1970, organized interest groups have become more important in local politics. Local leaders report that all types of organization have become more active; this tendency has been particularly strong in municipalities experiencing fiscal problems (Mouritzen, 1985a, p. 28). Table 4.6 presents the reports of different types of local organizations concerning the use of a number of strategies.[11]

TABLE 4.6

Selected Activities and Representation of Local Interest Groups (in percentages) (N = 393-404)

	Labor Unions	Employers	Homeowners	Renters' Associations	Agriculture	Commerce, Service	Sport, Culture	Consumers	Idea-Based
Direct contacts with council members	68	73	61	40	30	39	31	28	52
Cooperation with parties	83	27	9	29	30	18	3	3	36
Political campaigns	54	18	9	11	13	7	3	24	52
Influence budget process	71	42	50	16	25	42	61	27	46
Average of ten activities	67	52	47	38	35	45	42	31	55
Direct representation in city council	36	8	20	22	8	22	3	3	0

SOURCE: Villadsen (1985a, pp. 129, 133).
NOTE: For some of the interest groups the number of respondents is rather low, in particular for employers' associations, where the relative frequencies are based on the reported activities of only 12 respondents.

According to the figures in this table, labor unions are the most active, and consumer groups and farmers' associations (agriculture) are the least active. When it comes to the selected activities, some striking differences appear. The traditional close cooperation between labor unions and the Social Democratic party is reflected in the importance of cooperation with political parties for this category of interest organization — 83% of all labor unions report having cooperated with political parties within the last year and one-third report that they are directly represented in the city council.[12] In contrast, neither employers' and farmers' associations nor commerce have the same close relations to a political party. What is perhaps somewhat surprising is the fact that farmers' associations are not, according to their own reports, represented in the city council.

Responses to Fiscal Stress

When fiscal stress hits a Danish municipality, it follows from the preceding description that local political leaders respond under severe constraints. These constraints, which originate in central government, in the political party, and in local interest politics, manifest themselves in *political costs* attached to various possible courses of action. The political costs vary between revenue-raising strategies and expenditure-reduction strategies; there also may be differences between municipalities due to particular socioeconomic and political constellations, and finally between the different programs that are the responsibility of a Danish municipality.

Increase Revenues or Cut Expenditures?

It was postulated above that expenditure-cutting responses to fiscal stress would generally produce higher political costs than would revenue-increasing responses. This hypothesis gains credibility if one looks at aggregate figures on actual fiscal behavior (Table 4.7). During 1978-1986, when Danish local governments had trouble meeting the previous year's level of expenditures without raising taxes (that is, they were hit by fiscal stress), we find a picture of increasing expenditures and taxes. As mentioned above, total expenditures would have dropped by almost 9% over the period with a constant rate of taxation and fixed fees and charges. However, as Table 4.7 shows, total expenditures have gone up by 9% during this period. Operating expenditures are 24% higher than

TABLE 4.7

Trends in Fiscal Behavior, Danish Local Government, 1978-1986

	1978	1979	1980	1981	1982	1983	1984	1985	1986
Operating expenditures	100	106	113	118	122	119	119	125	124
Capital expenditures	100	102	79	65	51	36	41	49	46
Total expenditures	100	105	106	108	109	103	103	110	109
Tax revenues	100	111	116	114	114	119	121	127	129
Fees and user charges	100	103	104	105	109	110	110	121	108
Total revenues	100	106	107	108	110	104	104	111	108
Tax effort	100	108	111	110	109	112	113	112	115

SOURCE: FAUI Project comparative data base
NOTE: All figures are per capita, constant prices. Expenditures are net (excluding conditional grants but not excluding fees and user charges). Total revenues include general grants but not conditional grants.

in 1978. To the extent that major expenditure reductions have taken place, this has clearly hit capital expenditures most severely.

Up until 1982, general grants increased; since then they have been reduced drastically, by approximately 70% (see Table 4.2). This reduction has been compensated for more or less through increased tax revenues. This compensation has come about through political decisions to raise the rate of taxation (tax effort increases from 109 in 1982 to 115 in 1986) and increases in the tax base (about 7% in real terms; see Table 4.2). What is important to notice, however, is the fact that improvements in the tax base are almost exclusively due to the fact that local government price/wage increases have lagged behind the rate of inflation. This is illustrated in Figure 4.2, which clearly shows that the cost of a given amount of municipal services has, for four consecutive years, dropped relative to the tax base. Therefore the value of the tax base in constant terms has increased. The improvement in real terms of the municipal tax base is primarily due to a relative decline in the salaries of municipal employees. Without central government intervention in the public labor market and without a centralized system of wage settlements, the fiscal crises would have been felt much more acutely by the local governments.

Also, survey data tend to confirm the image of the political leader as an actor who chooses the "strategies of least resistance." In Table 4.8

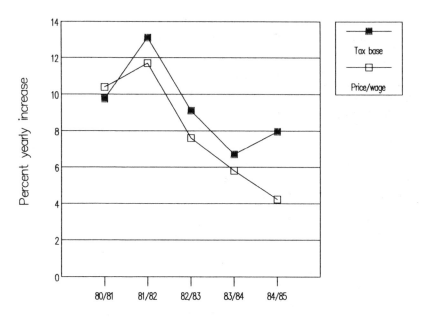

Figure 4.2 Increases in Tax Base and Prices and Wages, 1980-1985

various fiscal austerity strategies are ranked according to their perceived importance.[13] Note that the first four strategies either hit generally and marginally (increase taxes), are invisible (use liquid assets), or
hit with uncertainty; thus it is impossible to point to any individual or
organized group who will be hit by reduced capital expenditures or
deferred maintenance of capital stock.[14] Less used are reductions in
services (with little difference between services funded by own
revenues and by intergovernmental grants). The direct layoff of personnel is used infrequently, particularly for administrative personnel.[15]

City-Level Variations: Does Politics Matter?

It was indicated above (Table 4.3) that there were great variations
across cities in the seriousness of the fiscal crises. Also, it was suggested that partisan cleavages were generally important in Danish local
politics. Political parties gain support from different categories and
social classes that have different and sometimes opposing interests
when it comes to local fiscal policies. A major determinant of the

TABLE 4.8
Frequencies of Fiscal Austerity Strategies, Denmark (spring 1983)

Strategy	Index of Importance
Reduce capital expenditures	81
Increase taxes	72
Lower surpluses	71
Defer maintenance of capital stock	59
Freeze hiring	51
Increase user fees	51
Reduce services funded by intergovernment grants	44
Reduce services funded by own revenues	36
Increase long-term borrowing	35
Make purchasing agreements	33
Lay off service delivery personnel	26
Cut spending in all departments	25
Sell some assets	22
Add intergovernmental revenues	8
Contract out services to other units	8
Lay off administrative personnel	3
Reduce administrative expenditures	3

spending preferences of political leaders as well as citizens is party: Socialist parties prefer higher spending than conservative parties (Mouritzen, 1985b, 1987b). Also, it has been shown that political party is particularly important during times of fiscal stress (Mouritzen, 1985b, p. 11). It is therefore likely that there will exist differences between local governments in their responses to fiscal stress, differences that are a function of which political party holds power. More specifically, the inclination of a local government to cut expenditures is expected to be a function of the power of conservative parties as opposed to the Social Democratic and socialist parties.

A bivariate test of the party hypothesis does not produce supportive evidence — on the contrary. Operating expenses (in real terms) rose from 1982 to 1986 by 1.7% in municipalities with a Social Democratic mayor and 2.1% in municipalities with a mayor from one of the conservative parties. In fact, there is a significant *positive* relationship between the share of seats held by the conservative parties and expenditure growth ($r = .12$).

TABLE 4.9

Causes of Expenditure Changes, 1982-1986 (beta coefficients)

	Increase in Expenditures
Needs gap	.15**
Liquid assets, 1982	.20**
Change in needs, 1982-1986	.23**
Change in unemployed, 1981-1986	−.19**
Fiscal stress	−.28**
Social Democratic mayor	.09*
R^2	29.7

*p < .10; **p < .05.

In order to test the hypothesis in a multivariate setting, additional factors are introduced. It is further expected that changes in operating expenditures from 1982 to 1986 are a function of the following:[16]

(1) The size of a possible "needs gap" at the start of the period and changes in needs during the period: The larger the gap between needs and services and the larger the increase in needs, the larger the increase in expenditures.

(2) The stock of financial resources at the start of the period: The larger the stock, the larger the increase in expenditures.

(3) Changes in available resources over the period equivalent to fiscal stress: The larger the degree of fiscal stress, the smaller the increase in expenditures.[17]

The operationalizations of these factors are described in the Appendix to this chapter. The result of the test is found in the first column of Table 4.9.

Generally the expectations are confirmed. Expenditures increase fastest in municipalities where demands were highest and resources (liquid assets) in stock were greatest in 1982. Further, expenditures decline with decreasing needs and resources (fiscal stress). The effect of changes in the local labor market is somewhat surprising: An increase in the number of unemployed people is associated with a decrease in operating expenditures. Increasing unemployment notably does lead to increasing expenditure needs in policy areas such as social assistance. But it may also lead to a decrease in the need for other municipal services such as day-care institutions. The net effect of changes in unemployment evidently seems to be an increase in expenditures.

Finally, *party politics does matter.* All things being equal, local governments with a Social Democratic mayor do increase expenditures more than governments with a mayor from one of the conservative parties. In other words, the inclination to cut back is more pronounced in the latter group of municipalities.

Which Programs Are Hit? Four Models of Choice

Overall, total expenditures rose by close to 2% over the period 1982-1986. This fact does not imply that cutbacks did not take place. Many municipalities did operate at a lower level of activity at the end of the period (about 40% experienced a reduction in real expenditures from 1982 to 1986). Even in municipalities where aggregate spending was on the increase, cutbacks may have taken place in some programs. The question is which programs were particularly hit and which programs protected. Four analytic traditions suggest different answers to that question (see also Clark & Ferguson, 1983, pp. 18-20):

- *The central-local perspective:* According to this perspective, a program's sensitivity to expenditure cuts is a function of the degree of regulatory control. It can be expected that local political leaders will protect their "own" programs that have been built up during times of resource abundance via autonomous local decisions. This model therefore suggests that programs with greatest local autonomy are least likely to be cut.
- *The pressure group model:* Political leaders will seek to minimize political conflict and protest and to balance the interests of major groups. According to this model, we would expect cuts to take place in programs where the interest groups are least active.
- *Citizens' preferences:* According to this approach, fiscal policies are a direct expression of citizens' preferences. Politicians are therefore expected to spend more on areas where citizens want to expand, and less where they want to reduce spending.
- *Leaders' preferences:* According to this model, fiscal policymaking is a function of the preferences of local political leaders, independent of the activities of pressure groups and of the preferences of citizens.

The four models may complement each other, as they may potentially portray different aspects of or forces in the fiscal policymaking process. Clearly, a thorough test should be based on a city-level analysis. However, data for such a study are not immediately available, and therefore a first test of the four models will be based on aggregate data and on bivariate rank order correlations. The four models will be tested on survey data from the beginning of the 1980s, which makes it possible to rank ten programs according to (perceived) regulatory control, pressure group activity, and citizens' and leaders' preferences. These rankings are presented in Table 4.10.

The first column of this table repeats the ranking from Table 4.4 (public utilities and social counseling are left out due to data problems); that is, the ten programs are ranked according to the degree of regulatory controls (from low autonomy to high autonomy). In the second column, the programs are ranked according to strength of pressure group demands as perceived and reported by political leaders in the late fall of 1981.[18] The last two columns rank the ten programs according to the spending preferences of political leaders and citizens (from high to low spending dispositions). This ranking is based on the so-called percent-difference value, that is, the proportion of respondents who wanted their local government to spend less is subtracted from the proportion who wanted more money spent.

The rankings indicate that the preferences of citizens and political leaders are not correlated with organized pressure group activities; that is, groups do not mirror individuals' interest nor do the preferences of leaders reflect interest group activity. The best predictors of the strength of pressure group demand are the extent to which the service is divisible as opposed to public in nature and the existence of professions. There is a striking difference with respect to those services where group demands are among the strongest — day care, primary education, and collective transportation. On these three services politicians seem to hold moderate spending preferences. In contrast, roads, with relatively weak interest group demands, is the area in which political leaders want to expand the most. What is particularly interesting is a very strong *negative* relation between the degree of regulatory control and the spending preferences of political leaders: The more tightly a program is controlled by central government, the less popular it is among local political leaders.[19]

The dependent variable is the rank presented in Table 4.11, based on changes in local spending from 1982 to 1986. Impressionistic reading of this table and a comparison with the ranks in the first and second columns of Table 4.10 suggest that degree of regulatory control and pressure group activity are not at all related to development in program appropriations. This is supported by the rank order correlations of Table 4.12.

TABLE 4.10

Regulatory Control, Pressure Group Demands, and Preferences of Local Political Elite and Citizens, 1981

Regulatory Control	Pressure Group Demands	Preferences of Political Elite	Preferences of Citizens[a]
Social assistance	Sport, culture	Roads	Unemployment programs
Primary education	Day-care institutions	Programs for elderly	Programs for elderly
Unemployment programs	Primary education	Sport, culture	Collective transport
Administration	Collective transport	Unemployment programs	Day-care institutions
Collective transport	Unemployment programs	Libraries	Primary education
Day-care institutions	Programs for elderly	Day-care institutions	Sport facilities
Programs for elderly	Libraries	Primary education	Libraries
Libraries	Administration	Collective transport	Social assistance
Sport, culture	Social assistance	Administration	
Roads	Roads	Social assistance	

NOTE: The elite survey conducted in November 1981 involved 249 respondents (mayors, members of finance committees, and chairs of standing committees) in 40 municipalities. The survey of citizens conducted in December 1981 involved 1,020 citizens drawn from 104 municipalities. The rankings based on preferences (third and fourth columns) are based on a series of items where respondents were asked if they would prefer more spending, less spending, or if they were satisfied with present spending. The ranking is based on the proportion of respondents who preferred more spending minus the proportion who wanted less spending (percent-difference scores).

TABLE 4.11

Expenditures on Selected Functions, 1982-1986 (index, current prices)

	1982	1983	1984	1985	1986
Unemployment programs	100	146	150	154	138
Programs for elderly[a]	100	108	114	123	129
Administration	100	106	112	121	126
Day-care institutions	100	105	112	118	124
Sport, culture	100	105	112	118	124
Libraries	100	105	108	114	119
Roads	100	97	108	113	116
Primary schools	100	104	106	111	114
Collective transport	100	104	102	109	113
Social assistance	100	103	108	115	110

SOURCE: Statistiske efterretninger and Kommunalstatistiske meddelelser from relevant years.
NOTE: Expenditures are gross; that is, they include matching grants. Deflation figures are not available at this disaggregated level of analysis. The municipal wage/price index rose 21.5% over the period.
a. Expenditures for elderly programs do not include the cost of visiting home nursing, rent support, or pensions.

TABLE 4.12

Rank-Order Correlations Between Regulatory Control, Pressure Group Activity, Preferences of Politicians and Citizens, and Changes in Appropriation, 1982-1986

	Kendall's Tau
Degree of regulatory control	−.066
Pressure group activities	.156
Preferences of political elite	.333
Preferences of citizens	.571

NOTE: The first three coefficients are based on ten cases (programs); the last coefficient is based on eight cases.

Programs where pressure was most intense — sport and cultural programs, day care, and primary education — seem to lie in the middle when it comes to changes in spending. The first have experienced a slight increase in expenditures, while schools seem to have been the object of some reductions. In contrast, programs where pressure could be characterized as moderate — unemployment and elderly programs — score very high when it comes to increases in expenditures. What are hidden behind these figures, however, are major demographic changes that have occurred during the four-year period. The number of elderly people has increased by almost 6%, the number of schoolchildren has dropped by more than 8%, and the number of children eligible for day care has decreased by almost 13%. If these demographic changes are interpreted as changes in demand, what has happened in areas such as schools and programs for the elderly is simply that expenditures have changed with demand, while day-care appropriations per child have increased in real terms, probably around 15%.

The rankings of citizens' and politicians' preferences are somewhat correlated with the ranking of programs based on changes in appropriations. In particular, citizens' preferences seem to have some predictive power. The way the program mix in the local budgets changed during the period from 1982 to 1986 is therefore, to a very large extent, a reflection of the preferences of the population.

Conclusions

For some years now, Danish local governments have experienced fiscal stress, a situation where the budget could not be balanced without expenditure cuts or an increasing exploitation of the various resource bases, most notably the local income tax base.

Central government has played an important role in this period. On the one hand it is part of the problem, as it has reduced, over a four-year period, general grants by 70%, equivalent to about 10% of total local revenues. On the other hand, central policies are also part of the solution. The economic policies of the conservative coalition government have brought the Danish economy from a recession to a boom period (at least that is what the government has told the local governments), with positive consequences for the local tax base. Second, central government has, through the centralized system of wage negotiations, been able to implement a reduction (in absolute as well as in relative terms) in the salaries of public employees, including the work force of local governments.

Through its policies toward the local governments, central government has been able to solve its own fiscal problems. The deficit in the national budget has been reduced to zero in a four-year period, to a very large extent as a consequence of the reductions in grants to local governments.

In the short term, the policies of the central government can therefore be called a success. However, through these policies a couple of potentially very strong "bombs" have been created that can threaten fiscal policymaking in the coming years — in local as well as central government. As a consequence of the wage restraints and reductions in real take-home pay, public employees' unions have formulated demands for very large increases in salaries. These demands, the subject of negotiations in the spring of 1987, come at the start of a period in which — according to OECD predictions — Denmark is likely to enter a new recessionary period. This recession obviously will affect the tax base of local (and central) government. The combination of recession and growing demands of municipal employees is likely to lead to a new period of fiscal stress in local governments. The question is whether the responses from the previous period — 1982 to 1986 — will be repeated: Can we expect expenditures to be maintained and taxes to be raised? The answer depends on the extent to which the results of the analysis can be generalized over time.

It is difficult to point toward a single dominant factor in the fiscal policymaking process of Danish local government. The results of this study, however, suggest that policymaking has a strong local base and that it is — as proposed — biased in favor of public expenditures. It corresponds rather closely to citizens' preferences and to a lesser degree to the preferences of local political leaders. The intensity of group pressure as perceived by local political leaders does not correlate with changes in program appropriations. If one takes into consideration changes in "objective demand" as measured by demographic shifts, interest groups may, however, have been important, but the exact effect cannot be revealed in this kind of analysis. Finally, the extent of regulatory controls is not correlated significantly to local priority setting.

There is no doubt that changes in fiscal policies in the period from 1982 to 1986 are mainly linked to the local political and socioeconomic environment of local political leaders and that the configuration of political parties does matter. It seems as if expenditures are linked more to resources in bourgeois municipalities and more to social needs in municipalities where the Social Democratic party holds power. The analysis does not confirm the observation, often made, that fiscal stress leads to centralization.

There is no reason the same mechanisms should not be at work in the future. On the contrary, as large groups of "dependents," that is, clients and employees of local governments, perceive of themselves as the real losers in the last four years of conservative government, so much dissatisfaction accumulates that the political costs involved in major expenditure reductions are still likely to be very high.

Appendix

Most of the data used for this chapter were collected as part of the project on Central Control, Local Communities and Local Politics, which was carried out between 1980 and 1985 with support from the Danish Social Science Research Council and the Municipal VAT-Fund. The chapter draws on several surveys conducted by the project in the early 1980s in a stratified sample of 40 Danish municipalities:

- *political elite survey:* conducted as an enquete in November 1981, covering 249 politicians, mayors, members of the finance committee, and chairmen of the standing committees; response rate 75%
- *administrative elite survey:* interviews in the spring of 1982, covering the city director and heads of the budget office, the social and health department, and the educational department; response rate 98%
- *budget officer survey:* conducted in the spring of 1983 (enquete) as a follow-up on the 1982 interviews; survey instruments partly based on those developed in the FAUI Project; response rate 80%
- *party survey:* conducted in 1982-1983 as an enquete, covering all party organizations in the 40 municipalities; N = 337; response rate 60%
- *survey of local organizations:* conducted in 1982-1983 as an enquete, covering selected types of organizations in the 40 municipalities; N = 433

The dependent variable of the regression model of Table 4.9 is operationalized as follows:

- *increase in expenditures:* percentage increase in operating expenditures (net of conditional grants) from 1982 (final accounts) to 1986 (budget) measured in constant prices

The independent variables are operationalized as follows:

- *Needs gap* is calculated by means of the coefficients revealed in an analysis of citizens' spending preferences (see Mouritzen, 1987b). In this analysis about 20 factors were found to affect the preferences for aggregate spending of individual citizens — some were individual characteristics, others community characteristics. At this point I have not had access to data that would allow for a comprehensive simulation of the preferences of citizens in every municipality. Instead, two policy measures — service level in 1982 and level of taxation in 1982 — were used, both of which had significant effects on the spending dispositions of citizens. Both variables were standardized and weighted with the beta scores found in the analysis of individuals' preferences.

- *Liquid assets* is total liquid assets per capita 1982 divided by population 1982.
- *Change in expenditure needs* is measured as needs in 1986 divided by needs in 1982. In the assessment of needs, the age categories and corresponding weights from the formula for distribution of grants to the municipalities were used.
- *Change in unemployed* is measured as percentage change in number of unemployed in the municipality (full-time equivalents).
- *The fiscal stress measure* (also used in Table 4.3) is calculated as the percentage by which operating expenditures (in fixed prices) would have been cut if (1) capital expenditures are kept constant, (2) tax rates are kept constant, (3) fees and user charges are unchanged (in fixed prices), and (4) liquid assets are not used or built up. The measure therefore mainly reflects changes in taxable income, general grants, and the extent of deficit financing in 1982.

The models in Table 4.9 were carefully examined for multicollinearity. The correlations between the independent variables as well as the tolerance levels revealed at each step of the regressions clearly showed that multicollinearity is *not* a serious problem. The final models were evaluated by means of probit plots that showed that residuals were normally distributed.

Notes

1. The time series in this and the following tables have been collected as part of a comparative project (see Mouritzen, 1987c). All figures exclude county government and the two metropolitan jurisdictions of Copenhagen and Frederiksberg. In the chapter, analyses of expenditures and revenues exclude conditional grants; that is, expenditures are counted as net (including fees and user charges). Exceptions from this are Table 4.11 and Figure 4.1, where conditional grants (excluding reimbursements for pensions and children's allowances) are included. All calculations are based on fixed prices using the special municipal wage/price deflator. Figures from 1978 to 1985 are based on final accounts, while 1986 figures are based on budgets.

2. The use of liquid assets to close the revenue/expenditure gap is a normal and completely legal strategy in times of fiscal stress (and was in fact used every year from 1979 to 1982). As liquid assets are limited, they are not, however, counted as a revenue source in the calculation of the fiscal stress measure.

3. The definition of fiscal stress explicitly mentioned "real expenditures." In the figures shown in Table 4.1, changes in expenditure needs are not considered (except for population changes). It is therefore important to notice that the time series probably underestimate the extent of the fiscal crises in the beginning of the decade because a combination of mandated costs and rising unemployment forced most municipalities to increase expenditures just in order the maintain service levels. The extent of this phenomenon is impossible to assert.

4. Changes in the value of the local tax base are to a large extent a function of changes in the salaries of municipal employees (see below).

5. Notice a slight difference in this operationalization of fiscal stress from Table 4.1. Here we look at operating expenditures and assume therefore that capital expenses are kept constant.

6. Also, the number of counties was reduced to 14 and a two-level system was created with a relatively clear demarcation of functions between the municipalities and counties.

7. The one exception is the metropolitan area, which is still divided up among many jurisdictions.

8. There are surprisingly great similarities between the perceptions of city directors and the perceptions of the heads of the educational and health and social affairs departments.

9. As mentioned earlier, the grant system has a strong equalizing effect. When, in the beginning of the 1980s, central government started to consider grant reductions as a viable strategy to solve its own fiscal problems, there was a problem in that the degree of equalization was a function of the *amount* of grant money. Therefore in 1983 the system was changed so that equalization came to rest on intermunicipal transfers, leaving central government free to establish a grant policy independent of equalization objectives.

10. Social Democratic local politicians had approached fellow partisans in government or parliament twice or more in 83 of the responding municipalities (since the last election) in order to bring up, promote, or change an issue concerning their local government. Less used was the party channel of influence for the Liberal party, while the Conservative party and the other right-wing parties did not use the party channel very often (Mouritzen, 1983, p. 12).

11. This was taken from a 1982-1983 survey of local interest organizations; see the Appendix to this chapter.

12. This is according to their own reporting. The question asked whether or not they felt they were directly represented (as an organization) in the city council.

13. This is from a survey with heads of budget offices in the spring of 1983; see the Appendix to this chapter. The index of importance is calculated on the basis of the importance attached to each strategy. The categories — very important, somewhat important, little importance, and not used — were assigned weights from 3 to 0, the relative frequencies were multiplied by the weights, and the sum was subsequently standardized to an index ranging from 0 to 100.

14. The major parts of construction and maintenance are contracted out to private business; it is therefore uncertain what firms and what individuals will be hit if expenditures for these purposes are reduced.

15. These perceptions are, to some extent, a result of the particular time of the survey. The data were collected 6 months after the first round of grant reductions that were decided *after* the 1983 budget was passed. Therefore local government could not meet the resulting gap by increasing taxes. The only solutions left were expenditure reductions, hiring freezes, use of liquid assets, and the like. Therefore, the data possibly exhibit a serious bias in *disfavor* of the predictions made. So the general argument really finds support in the patterns discerned.

16. This is a brief and somewhat simplified presentation of some of the arguments presented in Mouritzen (1987a, pp. 7-8).

17. Of course, the fiscal behavior of a local government is also likely to be influenced by the particular interest configuration in the local community. This factor is left out of the models estimated, as data were not available for the 273 municipalities.

18. The ranking of group pressure is based on the proportion of respondents (N = 249, excluding unknowns) who reported demand to be strong or moderate. The scores range from 85% on sport and culture to 35% on roads.

19. Tau equals −.644 and is the only significant relation found between the four explanatory rankings.

References

Bentzon, K.-H. (1981). *Kommunalpolitikerne.* Copenhagen: Samfundsvidenskabeligt Forlag.

Bruun, F., & Skovsgaard, C. J. (1981). Local determinants and central control of municipal finance: The affluent local authorities of Denmark. In L. J. Sharpe (Ed.), *The local fiscal crises in Western Europe.* London: Sage.

Clark, T. C., & Ferguson, L. C. (1983). *City money: Political processes, fiscal strain, and retrenchment.* New York: Columbia University Press.

Dilling Hansen, M. (1983). *Marginaler paa kommunaldirektoerpoergeskema* (Report No. 17). Odense: University of Odense, Central Control, Local Communities and Local Politics.

Mouritzen, P. E. (1983). *Modelling central-local relations: Six models for the study of the interplay between central government and local levels of government* (Report No. 26). Odense: University of Odense, Central Control, Local Communities and Local Politics, Institute of Public Finance and Policy.

Mouritzen, P. E. (1985a). Concepts and consequences of fiscal strain. In T. N. Clark, G.-M. Hellstern, & G. Martinotti (Eds.), *Urban innovations as response to urban fiscal strain* (pp. 13-33). Berlin: Verlag Europäische Perspektiven.

Mouritzen, P. E. (1985b). Local resource allocation: Partisan politics or sector politics. In T. N. Clark (Ed.), *Research in urban policy* (Vol. 1, pp. 3-19). Greenwich, CT: JAI.

Mouritzen, P. E. (1987a). Dimensions of fiscal strain. In T. N. Clark (Ed.), *Research in urban policy* (Vol. 3). Greenwich, CT: JAI.

Mouritzen, P. E. (1987b). The demanding citizen: Driven by policy, self-interest or ideology? *European Journal of Political Research.*

Mouritzen, P. E. (1987c). *Consolidation and autonomy: Fourteen propositions about fiscal stress in local government.* Unpublished manuscript.

Villadsen, S. (1985a). Partierne i lokalpolitikken: Statsliggoerelse, politisering og lokal indflydelse. In S. Villadsen (Ed.), *Lokalpolitisk organisering.* Copenhagen: Jurist- og Oekonomforbundets Forlag.

Villadsen, S. (1985b). Lokal korporatisme? Organisationerns og bevaegelsers rolle i den lokale velfaerdsstat. In S. Villadsen (Ed.), *Lokalpolitisk organisering.* Copenhagen: Jurist- og Oekonomforbundets Forlag.

Wolman, H., & Davis, B. (1980). *Local government strategies to cope with fiscal stress.* Washington, DC: Urban Institute.

5

Local Leadership and Bureaucratic Legitimacy: The Finnish Case

ARI YLÖNEN

> The rational law of the modern occidental state, on the basis of which the trained official renders his decisions, arose on its formal side, though not as its content, out of Roman law. The latter was to begin with a product of the Roman city state, which never witnessed the dominion of democracy and its justice in the same form as the Greek city. (Weber, 1927/1987, p. 339)

European research on local politics and leadership has been flourishing in recent years. This may sound surprising because in most European countries local governments are gradually losing fiscal independence relative to central government. However, as the public sector ceases to expand and the need to minimize expenses becomes increasingly evident, most of the problems caused by these trends visibly affect cities. Local government, for example, is the visible element in the allocation of welfare services through its responsibilities for service delivery. And new social movements no longer based on traditional divisions of social classes or parties are active locally. Accordingly, there are several reasons for researchers to become increasingly interested in studying local politics (Clark & Ferguson, 1983; Cochrane, 1983).

The Fiscal Austerity and Urban Innovation Project studies fiscal strategies preferred by local governments in this new context. The FAUI Project in Finland has two types of objectives: those that are significant from a country-specific view and those that enhance international comparisons. The main question concerning Finnish municipal administration is whether local decision makers adopt principles of new decision-making cultures, structures, and legitimation routines that differ significantly from those used by national decision makers. The study is based on the hypothesis that local Finnish leaders seek legitimacy for fiscal decision making from bureaucratic authority. That is, local leaders seek legitimacy for their strategy preferences by functioning as the representatives of the public sector, as civil servants. Thus, perhaps, they do not face as strong local political pressure to seek legitimacy from individual citizens, from pressure groups, or from party organizations as in other countries. (See Appendix A, this chapter.)

Why the argument that local leaders in Finland seek decisional legitimacy from bureaucratic authority? According to Max Weber, bureaucracy is the modern form of a developing capitalistic structure and a social organization demanded by changing production means. As Andreski (1983) interprets Weber, "To develop, industrial capitalism required an elaborate legal code regularly and predictably administered. For this reason, it seems, a legal system which exhibits these features is labelled by Weber 'rational' and a development in this direction is described as 'rationalisation of the law' " (p. 10). Weber (1927/1987) sees the rational state as based on expert officialdom and rational law and as essential to the growth of modern capitalism (p. 339).

The Fiscal Austerity and Urban Innovation Project in Finland

The Finnish study is based on the following assumptions:

(1) Politically, there is a nearly unanimous wish to decentralize administrative power. However, simultaneously, the arena for fiscal decision making is moving away from local communities, even extending to the level of international economic systems. This is the dilemma of local autonomy: a contradiction between generally accepted political aims and ongoing economic trends. It is often difficult to acknowledge this dilemma because of the great wish for more local autonomy.

(2) There are many reasons to assume that some characteristics of decision-making practice and structures increasingly apply to all levels of public administration, from the state to the local level. Some researchers see changes in national decision-making practices as a change in the decision-making culture. Many changes in the national decision-making culture also are adopted at the local level (Heiskanen, 1977; Villadsen, 1986).

(3) On the basis of previous studies, it is presumed that the social background and attitudes of local leaders are of minor significance in explaining choices between different policy strategies. But differences between leaders and leadership characteristics become more significant in studying how local leaders respond to outside impacts (e.g., state government, new trends in planning ideas) and where they seek legitimacy for new decision-making practices (Jacobs, 1971). From a sociological point of view, it is important to know to what extent preference profiles and legitimation processes become well established — that is, to what extent they become institutionalized.

In analyzing the changing relationships between municipalities and the state, attention is directed to those mechanisms used to carry out common objectives for the entire public sector. Mouritzen and Narver's (1986) description of Denmark is valid also for Finland, although Finland relies on fiscal incentives, not consolidation of municipalities, for coordination:

Local governments in Denmark are consolidated and they provide a comprehensive, universal system of services. When fiscal stress strikes, its effects are muted by a set of legal and political mechanisms that ensure stability both for governments and for individuals. Political leaders have considerable freedom in developing policies and implementing programs that are consistent with national party goals, local planning and preferences of citizens. (p. 212)

Cultural Heritage, Leadership Types, and Legitimacy Sources

Finland is small, one of the most sparsely populated countries in Europe, and relatively young as an independent nation. Concentration of the population into villages or larger centers has not occurred to the extent elsewhere in Europe. At present, about 75% of the population resides in centers; the remainder is dispersed. Industrialization reached Finland at a very late stage, and its influence on urbanization was generally limited. By 1920, 16% of the population lived in towns. In the 1940s, the population of towns began slowly to increase (National Housing Board, 1987, p. 10). But urbanization still was not keeping pace with the changing industrial structure. This could be seen as a policy to conserve Finland as an agricultural society as long as possible; as late as the 1950s, for example, attempts were made to stimulate small-scale farming. Finland was one of the few OECD countries where the absolute number of farms still grew in the 1950s (Ylönen, 1985, pp. 294-295).

There are many reasons to assume that the geographical size of the nation and late urbanization are relevant in this context. Even before European economic integration, the international relations of small nations were dependent on the fact that their own markets were not large enough to guarantee steady economic growth. Because of a small domestic market, a small nation cannot produce everything that is needed or use everything that is produced. Liberal international trade relations are not a political choice but an economic necessity. On the other hand, according to Eisenstadt (1985), small European states are the most "successful" ones in their spatial organization, that is, in their structure of centers and their center-periphery relations. According to Eisenstadt, the most important characteristics of their political structure are as follows:

(a) A high degree of concentration of the decision-making powers in the hands of the executive and/or bureaucracy with comparatively limited scope for open parliamentary conduct of affairs.
(b) An emphasis in internal arrangements on "allocative/distributive" policies and relations between different sectors, compared to the ideological or class distinctions more prevalent in the larger states.
(c) A relatively great importance of vertical multifaceted sectors or at least continuous cross-cutting among the more vertical segments, as against "class-conscious" socio-political groupings in the larger states.
(d) The combination of a comparatively high degree of segregation between external and internal issues in the actual policymaking process, and a concentration of decision making, both internal and external, in the hands of strong executive and/or bureaucratic organizations. (pp. 45-46)

The institutional density of social relations and the special institutional mechanisms and symbolic premises associated with this density are typical of small countries. Eisenstadt (1985) analyzes these as a combination of a very strong allocative and cooptational mode of internal political arrangements together with a strong emphasis on more open, principled, but symbolic politics, evident, for instance, in referenda or even elections. "Cooptation occurs largely through continuous interaction between the executive, the bureaucracy, the parliamentary commissions and the relevant interest groups, and only rarely in the parliamentary plenum" (pp. 47-48).

The National Context
of Local Democracy in Finland

The Early History

Although it is argued that Scandinavian countries followed a common "Scandinavian route" from the era of semifeudal societies and
absolutist states to similar contemporary societies with democratic
regimes and comprehensive welfare arrangements, certain features are
specific to Finland's development. Its path has definitely been more
uneven and more dramatic than that of other Scandinavian countries
(Alapuro, 1982, pp. 145-153).

For a long time, Finland was on the periphery of two larger nations,
first Sweden (until 1809) and then Russia. This era of political dependence on a great feudal state and as a periphery of another nation,
Sweden, was followed by nationalism as a mode of self-assertion and
liberation (Alapuro, 1982, pp. 118-119). While still an autonomous
grand duchy under the Russian Empire, Finland started to build a state
apparatus of its own in response to rising nationalistic tendencies. Thus
the development of a formal bureaucracy did not arise from industrialization and the confrontations based on it; rather, the development of
administrative machinery and the need to establish a sense of obligation to the Finnish state both took place at the same historical period.
After gaining independence in 1917, the young nation first suffered
civil war in 1918-1919 and then a threat of rightist agrarian takeover in
the early 1930s. After these dramatic events, Finland drifted into war
against the Soviet Union and was thus one of the many fronts of World
War II.

The First Period of Consensus Policies:
Central-Local and Public-Private Relationships

Surprisingly soon after gaining independence in 1917, Finland began
to amend its production structure by establishing economic units of
nationwide interest owned by the public sector. Agriculture was still the
most important source of livelihood, but the government started industrial production of its own and also began to support private industry. The year 1918 was thus the beginning of a period of many years
during which state-managed industrial policy was initiated through
purchases and investments — although on a small scale. Government
participation in national energy production was also agreed on without
political dissent; in 1918 an agreement was concluded giving the state

control over several limited companies. A year later the government gained controlling interest in these companies; along with control of these companies, the government got possession of large areas of forest land (Ahvenainen & Vartiainen, 1982). In addition, construction projects required by energy production were started between 1921 and 1924; a power plant serving a wide area was constructed and its electricity distribution network soon covered southern Finland.

It is understandable that the Finnish state became an industrial entrepreneur even though the general atmosphere in Europe emphasized freedom of competition. Having just gained independence, Finland wanted to ensure her economic independence. This had to be carried out by public financing because private and foreign enterprises were not willing to take the risk of investing capital in unstable conditions. Historians have described Finnish economic policy as follows:

> The economic policy of the young republic was based on the reserved liberalism of the latter half of the 19th century. Industrial and socio-political regulation and public enterprise (by the state and municipalities) were accepted to a certain extent but otherwise the basis was a market economy with private enterprise and freedom of activities in which factors of production were able to find the most suitable and useful places and in which the best result was thought to be achieved by the least possible government intervention. (Ahvenainen & Vartiainen, 1982, p. 178)

The relationship between the state and local administration is similarly a history of public-private relationships. At the turn of the century, local administrations were unwilling to take on responsibility for service functions that were deemed important by the state. Because of this, and because of the lack of professional staff in local administration, central government was compelled to start performing new services in some areas. In the beginning, it leaned on the help given by the private sector. Health care and social security, which started during the first decades of the nineteenth century, are examples of this.

It is essential to note that in Finland the basic features of local administration were adopted at a time when liberalism emphasized the importance of local autonomy. Local administration was considered an instrument for carrying out liberalism by encouraging citizens' participation and achievement of educational aims. In the spirit of liberalism, attempts were made to make local administration a balancing power between national government and individuals, while simultaneously putting stress on the principle of citizens' right of self-determination. The solution was a two-level structure of public administration that still prevails despite discussions about reinforcing administrative capacity at intermediate levels.

As the statutory duties of municipalities increased, a system for dividing expenditures between local government and the state emerged. The so-called government subsidy and aid system has developed gradually since the beginning of the nineteenth century. The aim has been to encourage municipalities to participate in duties related to health care and social security. The system has always been relatively incoherent and complicated, but the basic idea has been clear: Resources are controlled by central government but municipalities are free to choose their ways of coping with the tasks.

The painstaking early political upheavals and the early political history of compromise in relationships between the public and private sectors (because of the lack of practical means to stand as a young nation otherwise — that is, not having sufficient capital to safeguard national resources) form the historical basis of the so-called Finnish consensus politics. However, partly because of developments in Europe, local government in Finland also achieved an extensive degree of autonomy, including the right to collect local taxes.

The Second Phase of Consensus: The Welfare State Since the 1960s

In the industrial policies of the 1960s, the private sector still accepted industrial activity by the public sector as a solid part of the production structure. Multinational enterprise in Finland was rare; private Finnish industries obviously believed that the existence of public industry would automatically improve the condition of private firms. The term *mixed economy pattern* entered the vocabulary, referring to cooperation and mutual understanding between the private and the public sectors. Government did not aim at guiding the production structure in general, but there was room for public sector units alongside the private sector and they gradually formed an essential part of economic growth.

According to the statistics, the growth of public spending in Finland has been rapid since the 1960s, but not as rapid as in other Scandinavian countries (see Table 5.1). Services have been delegated from the private sector to the public one to such a great extent that private sector spending in 1976 was only 52% of the national product, whereas in 1848 it was 65%. The growth was fastest in health care and education.

TABLE 5.1

Public Expenditure as a Percentage of the GDP in the OECD Countries, 1960-1980

Country	1960	1970	1980
Denmark	24.8	40.2	56.2
Finland	26.7	31.3	38.4
Iceland	28.2	29.6	34.4
Norway	29.9	41.0	48.9
Sweden	31.1	43.7	62.1
France	34.6	38.9	46.4
Germany (West)	32.0	37.6	48.3
Italy	30.1	34.2	46.1
The Netherlands	33.7	45.5	59.5
United Kingdom	32.6	39.3	45.4
United States	27.8	32.2	35.0
Total OECD	28.5	32.6	40.2

SOURCE: Adapted from Kosonen (1987).

But the structure of public spending changed between 1975 and 1985, as shown in Table 5.2. Sociopolitical philosophy in Finland changed rapidly during the 1960s and the 1970s. New "super-ideas" were introduced, such as the idea of economic growth, the idea of expanding the welfare tasks of the public sector, the idea of regionally balanced social and economic growth, and the idea of the development of representative and participatory democracy (Heiskanen, 1977). Along with these, a new culture of national political decision making was agreed upon: an income policy decision-making system. This so-called liberal corporatist decision-making system is characterized by a system of conciliation between the representatives of unions and government in which all parties involved are compelled to debate their interests (Helander, 1982, pp. 163-168).

These rapid developments would not have been possible without a powerful national coalition. The Centrist party and the Social Democratic party ended up cooperating beyond the traditional limits of class division. A coalition with such heterogeneous supporting groups requires the support of a coherent ideology; the neutral "ideology of guidance" was the natural choice. This term refers to a public sector providing basic services and welfare; planning and improvement of these activities as well as of a nationally balanced service network are the cornerstones (Heiskanen, 1977, p. 74).

TABLE 5.2

Structure of Public Spending, 1975-1985 (in percentages)

	1975	1980	1985
General administration	10.2	10.3	9.3
Public order and safety	6.8	6.1	5.7
National defense	7.8	7.8	7.7
Education	27.6	26.9	24.9
Health care	21.0	21.4	22.6
Other social services	11.3	12.8	14.8
Activities serving traffic	6.6	5.4	5.0
Other purposes	8.7	9.3	10.0
Total	100.0	100.0	100.0
Public spending, of spending in total	23.7	25.1	27.3

SOURCE: Statistical Office of Finland (1986, p. 6).

Strong guidance by government can also be explained by the fact that rapid changes in industrial structure, expansion of the economy, and rapid urbanization did not take place until the postwar years. The metal industry, for example, was modernized because of the war-indemnity industry. "The frantic years" of urbanization took place relatively late, in the 1960s and 1970s. This situation was receptive to the establishment of many planning systems and can be seen as an era of public planning enthusiasm.

In briefly characterizing the changed relations between the state and municipalities, it is not an exaggeration to say that in a couple of decades, local administration came to perform welfare tasks defined at the national government level (see Table 5.3). When carrying out social policy aiming at a balance between different areas and segments of the population, municipalities are, in this respect, performing the duties of state administration. However, they also have been able to maintain some important rights that ensure a more autonomous status. One of the most important rights, the right to collect taxes, has resulted in competition between municipalities: They compete for affluent taxpayers and companies.

The situation is problematic. The framework of local fiscal activity is, to an ever increasing extent, defined by the economic aims of the entire public sector. Economic decision making is thus a part of a larger decision-making arena, influenced increasingly by international economic changes. Thus big and middle-sized towns, in particular, try to

TABLE 5.3

Expenditures of Urban Municipalities, 1980-1985 (in percentages)

	1980	1981	1982	1983	1984	1985
General administration	2.8	2.9	2.9	3.0	3.0	3.0
Public order	1.9	1.8	1.8	1.7	1.6	1.6
Public health care	10.4	10.5	11.3	11.7	13.0	13.2
Social welfare services	11.4	11.6	11.9	12.6	13.5	14.1
Education and culture	18.5	18.3	18.7	19.1	18.5	18.7
Community planning and public works	5.3	5.8	6.0	6.0	6.0	5.6
Real estate	4.2	4.3	4.2	4.1	4.1	4.0
Business enterprises	19.6	19.6	19.1	18.2	17.3	17.4
Inner service activity	2.0	2.1	2.1	2.1	2.1	2.1
Financing	3.1	2.8	2.6	2.4	2.5	2.7
Capital	20.8	20.3	19.4	19.1	18.4	17.6
Total expenditure	100.0	100.0	100.0	100.0	100.0	100.0
(1,000,000 FIM)	(28,294)	(32,842)	(37,168)	(40,522)	(46,161)	(52,282)

SOURCE: Statistical Yearbook of Finland (1982-1987).

adopt decision-making methods suitable for this competitive situation. Following similar trends in other countries, more formal planning and decision-making procedures are accompanied or replaced by new decision-making routines that enable local leaders to react flexibly to changes in economic competition. The consequent problem of democratic control is similar to that at the national government level, where interests of different parties are expressed and reconciled by top leaders in order to maintain international competitiveness. Thus, to be economically successful, local administrations must give leaders a negotiation mandate nearly up to the final decision. Yet at the same time, there is great pressure to decentralize administration and to increase local democracy.

Local Government and Municipal Functions in Finland

Although Finland as a nation is relatively young, the system of local government and the self-governing of towns are the results of a long historical process. Towns were self-governing as far back as the Middle

Ages, and the German Hanseatic system was also adopted in Finland. But the Finnish municipality in the modern sense was created in the nineteenth century. A statute enacted in 1865 made Finnish rural municipalities self-governing by separating ecclesiastical and secular administration at the local level. In the cities, a system of local government in the modern sense was implemented by statute in 1873.

The Finnish Constitution states that municipal administration must be based on self-government by the citizens. The most important features of local government are generally agreed upon as follows (Finnish Local Government, 1981, p. 7):

(1) Municipal authority is general and broadly based (a municipality can handle all matters of common interest that the law does not entrust to a special body, or that are not under the jurisdiction of another authority).

(2) Decisional power resides with persons elected by direct ballot.

(3) The state can assign new functions to the municipalities or take others away solely on the basis of law.

(4) Municipalities have the right to levy taxes.

Finnish citizens who are 18 years of age at the beginning of the election year are entitled to vote in municipal elections for council members every four years. Members of the municipal board, committees, and the boards of various municipal institutions are then generally chosen by the city council. Municipal democracy is thus representative in character. Furthermore, the 1976 Local Government Act states that municipal residents have the right to take initiative in matters concerning administration. A second important regulation in that act requires the municipality to provide the public with information on matters of general interest currently in the pipeline before they come to a vote. The fact that municipalities in Finland are self-governing does not, however, mean that their autonomy is unlimited; municipalities remain subject to state supervision and control.

The highest decision-making body in a municipality is the city council. The number of members elected depends on the municipality's population. The smallest council has 17 members, and the largest 85. In the Finnish municipal system, the council selects the city manager (also called the mayor); this person acts as a civil servant who is chosen for an indefinite period, usually by vote in the city council. She or he may be a member of the city council, but other candidates are possible. Depending on the political majority or type of political coalition in the

TABLE 5.4

Revenues of Urban Municipalities, 1980-1984 (in percentages)

	1980	*1981*	*1982*	*1983*	*1984*
Total of taxes and income of tax character	39.5	39.9	39.0	39.4	39.1
Shares, subsidies, and remunerations of state	14.2	14.0	14.6	15.4	16.5
Fees and fares	20.6	20.8	19.1	19.4	18.9
Loans	3.0	2.7	2.9	2.4	2.6
Other revenue	22.7	22.6	24.4	23.4	22.9
Total	100.0	100.0	100.0	100.0	100.0
(1,000,000 FIM)	(28,391)	(32,797)	(37,144)	(40,584)	(46,170)

SOURCE: Statistical Yearbook of Finland (1982-1987).

city council, the councillors as well as candidates for office recognize which party can put forward the most viable candidate.

The municipal board is appointed by the city council and tends to include most important local politicians, from nearly every party. The board has at least seven members; they are appointed, with their personal deputies, for two-year terms. The chair of the city board and one or two vice-chairs are chosen by the council from among the members. The city board meets weekly to carry out the decisions of the city council officially, as presented by the city manager, and to look after day-to-day administrative routines. Its responsibilities include planning and land policy, finances and accounting, and supervising the work of sector committees, and local government officers and employees. It also monitors municipality funds and assets and oversees the implementation of the budget. Informally, political agreement on most issues is reached at this level; since the members are broadly representative of every political group, the board offers a forum for testing the political feasibility of different policies.

Income tax paid by individuals is the most important source of revenue for municipalities. This tax is a given percentage of the taxable income of each taxpayer. In the 1980s the proportion of municipal expenditure covered by municipal tax revenues averages about 40% (see Table 5.4). Depending on the municipality, the tax rate currently varies from 14% to 19%; this decision is made by the city council.

Perceived Fiscal Problems

Finnish FAUI survey data were collected by mail in 1985. The survey was organized in cooperation with the Association of Finnish Cities and a letter of recommendation from the Association was enclosed with the questionnaire. The survey included all 84 municipalities officially classified as cities in 1985. Three groups of respondents — city managers, finance managers, and chairmen of city boards in each city — received identical questionnaires. The response rate for each group was as follows: city managers, 90.5%; finance managers, 95.2%; chairmen of city boards, 82.1%; first vice-chairmen, 79.8%; and second vice-chairmen, 75.8%.

These groups were selected because of the interest in the differences in local leaders' roles: City managers are expected to act as permanent civil servants, with, in most cases, a political affiliation; finance managers are expected to act as officials, and chairmen as politicians. In Finland, city managers are usually selected from among the candidates of different political parties, but in their capacity as managers they act as permanent officials. In fact, city managers are expected to fulfill the expectations of dual roles (civil servant and politician) although some of them have given up their party membership. Finance managers, too, are selected by the city council from among candidates acceptable to the political parties, although the parties do not officially nominate candidates. They are expected to support the multiparty management of the city as efficient officials. In contrast, chairmen of city boards are "pure" politicians; in most cases, they have full-time jobs outside the city government. They may later run for city manager (or deputy manager), which may affect their political orientation within the administration.

To devise a list of local fiscal problems, items from the U.S. FAUI Project CAO questionnaire were used and revised to fit Finnish circumstances. The main reason for reformulation is that cities in Finland have, during the past few decades, been generally financially stable and rather affluent. This stability is expected to continue due to the economic agreements between the state and the cities. Thus the respondents were not asked to look specifically at the three past years (as in the U.S. survey) but at local fiscal problems in general.

The officials' opinions concerning problems of city finance are analyzed in two ways. First, the problems seen as the most difficult by local leaders in different roles are explored. Second, the structures of

TABLE 5.5

Responses to Item: "How much do the following factors cause problems?"
(in percentages)

Very Much or Rather Much	Finance Managers (N = 80)	City Managers (N = 76)	Chairmen of City Boards (N = 183)
Reductions of state grants	78.9	78.2	87.4
Inflation	26.3	33.8	35.0
Unemployment	64.5	60.0	70.5
Reduction of tax revenues	75.0	85.0	77.0
City structure	47.4	61.3	49.7
Citizens' rising demands for services	69.7	71.3	59.6
Citizens' demands to avoid increase of local taxes	50.0	53.8	50.3
Difficulties in obtaining loans	53.9	43.8	47.0
Councilors' demands (and demands of other elected members of city government)	56.6	63.8	38.3
Top officials' demands	22.4	41.3	30.6
Other officials' demands	18.4	18.8	24.0
Industrial structure of the city	51.3	63.8	60.1
Decision making for economic development	30.3	30.0	45.4
Statutory tasks (according to laws covering sectors of public administration)	68.4	75.0	77.6
Delegating functions from the state to municipalities	76.3	85.0	86.9

perceived problems are examined for these three groups, as depicted through factor analysis. Both types of analysis show three types of problems, but the relationship with the state seems to be seen as most problematic, regardless of the type of leadership role (see Table 5.5). This reflects a state of affairs that has received a great deal of publicity in Finland. According to local leaders, new tasks given by the state and government subsidies given for the tasks are not balanced. Thus the internal bureaucracy of the public sector is not flawless. In addition, most local leaders see reduction of tax revenues, unemployment, and citizens' demands as problems for city finances.

Factor analyses show that although perceptions of the main problems are rather similar across leadership roles, there are differences in perceptions of how problems and pressures are mutually interrelated (see Table 5.6). The pressure on city managers and finance managers is evident: In their answers, local and delegated demands shape one factor

TABLE 5.6

Factor Analysis of Perceived Problems (principal axis factoring)

	City Managers (N = 76)			Finance Managers (N = 80)			Chairmen of City Boards (N = 183)		
	I	II	h^2	I	II	h^2	I	II	h^2
01	.22	.58	.39	.57	-.14	.34	.07	.47	.22
02	-.07	.32	.11	.65	-.05	.43	.27	.28	.15
03	.15	.54	.32	.59	.01	.34	.59	-.02	.35
04	.20	.86	.78	.48	-.21	.28	.60	-.03	.36
05	.00	.17	.03	.35	-.27	.20	.35	.11	.13
06	.37	.18	.17	.38	.21	.18	.05	.35	.13
07	.34	.22	.16	.47	.23	.27	.35	.23	.17
08	.13	.56	.33	.51	.05	.26	.38	.10	.15
09	.68	-.05	.47	.01	.52	.27	.08	.30	.10
10	.69	.12	.49	.01	.79	.62	.33	.27	.18
11	.55	-.07	.31	.23	.42	.23	.29	.29	.17
12	.30	.25	.15	.57	.20	.36	.65	-.08	.43
13	.40	.09	.17	.50	.14	.27	.32	.15	.12
14	.55	.04	.31	.41	.12	.18	.02	.68	.46
15	.65	.18	.45	.47	.12	.23	-.02	.71	.50
Eigenvalue	3.2	1.4		3.1	1.3		2.3	1.4	
Trace %	21.4	9.5		21.0	8.9		15.2	9.1	

and the financial means to cope with them constitute another. In contrast, chairmen of city boards see themselves as confronted more with social problems and worry about lacking state support to address them.

Pressure Groups and Formal Municipal Organization

As in most European countries, there are active local pressure groups in Finnish cities. Public interest in them is reflected in the large amount of space in the mass media devoted to their objectives and activities. In the last municipal election, in 1984, the Greens won a national share of 2% of seats in the city councils of Finland. But, in general, coalitions of big parties still run the city policies. Now, what is the role of parties vis-à-vis other sources of influence in municipal fiscal decision making in Finland?

TABLE 5.7
Most Influential Organizations or Bodies in Local Decision Making Concerning City Economy and General Decision Making[a]

Organization or Body	City Economy			General Decision Making		
	City Managers (N = 76)	Finance Managers (N = 80)	Chairmen of City Boards (N = 183)	City Managers (N = 76)	Finance Managers (N = 80)	Chairmen of City Boards (N = 183)
Pressure groups and organizations	22	–	–	22	–	–
Political parties	64	69	66	71	71	64
City council	84	78	81	84	79	79
City board	92	100	87	89	99	87
Top officials/city government	67	79	76	72	88	82
Top officials/city economic administration	50	68	46	28	45	34
State officials	42	43	39	36	43	32

NOTE: 100% = all persons from the group of answerers has given a score to the item (1-18) to be one among five. The items that have sum scores less that 20% have been left out of the table.
a. Percentage of sum scores in the list of five most influential bodies.

In the Finnish FAUI questionnaire, there was a list of 18 potentially influential groups, organizations, and bodies (see Appendix B, this chapter); groups not on the list could be added by respondents. Labor unions, civic groups, minority groups, mass media, political parties, commercial and business associations, and the like, as well as local formal decision-making bodies (city board, city council) and groups of local officials (e.g., top officials in the finance division or in central government) were included. All these are on the list both as sources of influence and as demands on the city government. Local leaders were asked to list, out of these 18 potential alternatives, the 5 most powerful ones in decision making about the city economy. Their perceived relative importance is an indicator of the most important sources of legitimacy in Finnish municipalities. Nearly 85% of the responses include the following, in order of importance: the city board, top officials in central city government, the city council, political parties in general, top officials in city finance division, and state officials. When the same question was asked about city government in general, the results were similar. All local leaders in different roles (city manager, finance manager, chairmen of city boards) agreed on this. Clearly, the executive branch of the municipality, top officials included, is the major source of influence in local government in Finland. Thus it is no wonder that, in Finnish municipalities, the most critical moments in party politics take place in the selection of local officials. Officials, too, carry their share of political responsibilities, and thus politics also plays an important role in the city council's selection of officials.

Preferred Fiscal Strategies

Data on fiscal strategy choices were collected according to the international FAUI study protocol. In the questionnaire, a list of 34 alternative fiscal strategies is used to collect two kinds of information from all three groups (see Appendix B): whether or not strategies have been adopted (the stage of adoption) and the importance of each strategy (see Table 5.8).

Drastic changes in fiscal strategy choices or preferences are not expected because of the fiscal stability of local economies and because of the institution of biannual economic agreements between the state and municipalities. These economic agreements are made between national municipal associations and the national government in order

TABLE 5.8

Importance of Strategies, Strategies That Already Have Been Adopted in Use, and Those That Have Been Considered for Adoption: Evaluation of City Managers (in percentages)

	Most Important	In Use	Under Consideration[a]
Revenues			
seek new revenue sources	26	X	
develop services to increase state grants			
increase local taxes			36 (51)
increase local taxes by furthering employment	72	XXX	
increase user fees and charges		XX	
sell real property (assets)			
defer some payments to the next year		X	
increase short-term borrowing			
increase long-term borrowing		X	
increase investments in stocks			
Expenditures			
make across-the-board cuts in all departments	57	XXX	
cut budgets of least efficient departments		X	
reduce expenditures in administration		X	
reduce expenditures for supplies and equipment		X	
eliminate some functions			43 (34)
extend the timetable for investments		XX	
eliminate some investments	34	XXX	
keep expenditure increases below rate of inflation			
make joint purchasing agreements			
Services			
increase cooperation between local governments		X	
reduce services funded by city's own revenues			42 (49)
reduce services funded by state grants			
improve productivity in services through better management	30	X	
improve productivity in services through labor-saving techniques	28	X	
contract with private sector for service provision		X	
improve productivity in services by increasing competition with private sector service providers			33 (41)
Personnel			
reduce personnel			34 (45)
reduce extra rewards for personnel			
keep down increases in wages and salaries	30	X	
postpone appointments in offices		X	
encourage early retirements			
deny overtime work			
encourage temporary take-off from duties without pay			
encourage extra holidays without salary			

NOTE: Only percentages for the seven most important strategies are given here. X = in use in over 50% of cities; XX = in use in over 75% of cities; XXX = in use in over 90% of cities.
a. In the same questions there are options: "is already being used" (= in use), "is already under consideration," and "is not yet been considered." Percentages for the five most often considered are shown here. In parentheses are percentages for "not yet being considered."

TABLE 5.9

Strategies Not Considered Important by Finnish City Managers, Finance
Managers, and Chairmen of City Boards (in percentages)

Strategy	City Managers (N = 76)	Finance Managers (N = 80)	Chairmen of City Boards (N = 183)
Increase short-term borrowing	71	75	54
Encourage extra holidays without salary	63	56	61
Deny overtime work	62	63	59
Encourage temporary take-off from duties without pay	59	56	56
Reduce services funded by state grants	55	64	61
Sell real property (assets)	49	60	66
Reduce extra rewards for personnel	47	48	47
Encourage early retirement	46	44	42
Increase investments in stocks	42	44	39
Increase local taxes	24	36	50
Reduce personnel	24	29	40

to achieve some common economic goals and levels of spending. They
include estimates of the level of inflation and increases in GDP as well
as recommendations for local government spending and revenue levels.
According to the constitution, local governments have the autonomy to
make their own fiscal and economic policy decisions; these economic
agreements are not binding, therefore, but in practice are taken into
account as far as possible. Therefore, for the Finnish case, it may be
more essential to study general orientations toward different types of
strategies and differences in opinions between local leadership roles.
To some extent, it is also possible to speculate on the near future by
studying which of the strategies not used at present are being considered for future adoption.

The strategy preferences shown in Table 5.8 and evaluations of strategies that are of least importance (Table 5.9) lead to three types of
conclusions. First, there is not yet much fiscal pressure for radical new
orientations in local economies. The most important strategies, much
more important than the rest, are administrative measures to raise more
revenues (by collecting income tax from an increasing number of
taxpayers) and to avoid waste in expenditures. Second, personnel

policy seems to be an arena of cautious stability; only the strategy to stop the increase of wages and salaries wins modest (30%) support from city managers.

Third, the strategies not currently in use but most often considered for possible future adoption have more opponents than supporters. These are the strategies to increase the level of taxation and to cut, reduce, or eliminate some services and functions. These do not seem likely without significant changes in Finnish local politics.

Conclusions

Nordic societies, not only in their present policy preferences but also by tradition, are characterized by a multitude of organized interests and by close cooperation between national organizations and public authorities. Due to constitutional provisions, communal legislation, and tradition, Finnish local leaders are well established as representatives and executors of communal autonomy. Their role, however, primarily is defined by an institutional system in which they have responsibilities for an extensive national system of services. This system encourages local leaders, and citizens as well, to act in order to achieve the national aims of the welfare society at the local level. But local leaders in Finland have the authority, depending on their expertise, to pursue national goals in a locally beneficial way.

From the point of view of local finances and local democracy, this system includes built-in paradoxes that demand solutions. Although an essential part of local administration activities involves carrying out statutory duties, competition among municipalities for resources is pervasive. One of the most popular ways for an average city to find new economic potential is to compete with other cities for firms and well-off taxpayers. To do so, the city administration adopts decision-making procedures suitable for this competitive situation; to be economically successful, local leaders need a strong negotiation mandate. The future challenge is to create sufficient flexibility in local administration to activate local resources — intellectual, organizational, and material — that are beyond the scope of state control, but to do so through open and democratic measures.

Appendix A:
Impact of Politicians and Officials
in Decision Making

TABLE 5.A1
Level of Expenditures in City Budget (in percentages)

	City Managers (N = 76)	Finance Managers (N = 80)	Chairmen of City Boards (N = 183)
Professional staff/very essential impact	32.9	18.8	24.0
Professional staff/rather essential impact	39.5	57.5	44.8
Equal impact	11.8	12.5	14.8
Councillors/rather essential impact	7.9	7.5	13.1
Councillors/very essential impact	5.3	2.5	2.7
Hard to say, no answer	2.6	1.3	0.5

TABLE 5.A2
Allocation of Expenditures Among Departments (in percentages)

	City Managers (N = 76)	Finance Managers (N = 80)	Chairmen of City Boards (N = 183)
Professional staff/very essential impact	21.1	11.3	21.3
Professional staff/rather essential impact	46.1	48.8	53.6
Equal impact	15.8	20.0	15.8
Councillors/rather essential impact	10.5	13.7	6.6
Councillors/very essential impact	3.9	0.0	2.2
Hard to say, no answer	2.6	6.3	0.0

TABLE 5.A3

Creation of New Strategies (fees, charges, new methods for saving, and so on)
(in percentages)

	City Managers (N = 76)	Finance Managers (N = 80)	Chairmen of City Boards (N = 183)
Professional staff/very essential impact	32.9	28.8	14.2
Professional staff/rather essential impact	39.5	52.5	41.0
Equal impact	13.2	7.5	29.5
Councillors/rather essential impact	6.6	5.0	9.3
Councillors/very essential impact	1.3	0.0	2.2
Hard to say, no answer	6.6	5.0	3.8

Appendix B

Organizations and Bodies

(1) professional staff and their unions

(2) unions as a whole

(3) homeowners' groups and organizations

(4) neighborhood groups and organizations

(5) civic groups, sport clubs, and youth organizations

(6) church and religious organizations

(7) minority groups

(8) taxpayers' associations

(9) commercial and business associations

(10) local media

(11) unions and groups for the elderly

(12) individual citizens

(13) political parties in general

(14) city council

(15) city board

(16) top officials in central city government

(17) top officials in city economic administration

(18) state officials

Strategies

Revenues

(1) Seek new revenue sources.

(2) Develop services that increase state grants.

(3) Increase local taxes.

(4) Increase local taxes by furthering employment.

(5) Increase user fees and charges.

(6) Sell real property (assets).

(7) Defer some payments to the next year.

(8) Increase short-term borrowing.

(9) Increase long-term borrowing.

(10) Increase investments in stocks.

Expenditures

(11) Impose across-the-board cuts in all departments.

(12) Cut budgets of least efficient departments.

(13) Reduce expenditures in administration.

(14) Reduce expenditures for supplies and equipment.

(15) Eliminate some functions.

(16) Extend the timetable for investments.

(17) Eliminate some investments.

(18) Keep expenditure increases below the rate of inflation.

(19) Make joint purchasing agreements.

Services

(20) Increase cooperation among local governments.

(21) Reduce services funded by city's own revenues.

(22) Reduce services funded by state grants.

(23) Improve productivity in services through better management.

(24) Improve productivity in services by adopting labor-saving techniques.

(25) Contract with private sector service producers.

(26) Improve productivity in services by increasing competition with private sector service producers.

Personnel

(27) Reduce personnel.

(28) Reduce extra rewards for personnel (overtime, travel, and other payments).

(29) Keep down increases in wages and salaries.

(30) Postpone appointments in offices.

(31) Encourage early retirement.

(32) Deny overtime work.

(33) Encourage temporary time off from duties without salary.

(34) Encourage extra holidays without salary.

References

Ahvenainen, J., & Vartiainen, H. J. (1982). Itsenäisen Suomen talouspolitiikka. In J. Ahvenainen, E. Pihkala, & V. Rasila (Eds.), *Suomen Taloushistoria 2*. Helsinki.

Alapuro, R. (1982). Finland: An interface periphery. In S. Rokkan & D. Urwin (Eds.), *The politics of territorial identity*. Beverly Hills, CA: Sage.

Anderson, J., Duncan, S., & Hudson, R. (Eds.). (1983). *Redundant spaces in cities and regions?* New York: Academic Press.

Andreski, S. (1983). *Max Weber on capitalism, bureaucracy and religion: A selection of texts*. London: Allen & Unwin.

Clark, T. N. (1985). Local government growth and retrenchment: Lessons from American cities. In T. N. Clark, G.-M. Hellstern, & G. Martinotti (Eds.), *Urban innovations as response to urban fiscal strain*. Berlin: Verlag Europäische Perspektiven.

Clark, T. N., & Ferguson, L. C. (1983). *City money: Political processes, fiscal strain, and retrenchment*. New York: Columbia University Press.

Clark, T. N., Hellstern, G.-M., & Martinotti, G. (Eds.). (1985). *Urban innovations as response to urban fiscal strain*. Berlin: Verlag Europäische Perspektiven.

Cochrane, A. (1983). Local economic policies: Trying to drain an ocean with a teaspoon. In J. Anderson, S. Duncan, & R. Hudson (Eds.), *Redundant spaces in cities and regions?* New York: Academic Press.

Eisenstadt, S. N. (1985). Reflections on centre-periphery relations and small European states. In R. Alapuro, M. Alestalo, E. Haavio-Mannila, & R. Väyrynen (Eds.), *Small states in comparative perspective*. Oslo: Norwegian University Press.

Finnish Local Government. (1981). *The Association of Finnish Cities and the Finnish Municipal Association*. Vantaa: Kunnallispaino.

Heiskanen, I. (1977). Julkinen, kollektiivinen ja markkinaperusteinen. Helsingin Yliopiston Yleisen Valtio-opin laitoksen tutkimuksia. Deta 31.

Helander, V. (1982). A liberal-corporatist sub-system in action: The incomes policy system in Finland. In G. Lembruch & P. C. Schmitter (Eds.), *Patterns of corporatist policy-making*. London: Sage.

Jacobs, P. E. (1971). *Values and the active community*. New York.

Kosonen, P. (1987). *Hyvinvointivaltion haasteet ja pohjoismaiset mallit*. Tampere: Vastapaino.

Mouritzen, P. E., & Narver, B. J. (1986). Fiscal stress in local government: Responses in Denmark and the United States. In T. N. Clark (Ed.), *Research in urban policy* (Vol. 2, pp. 197-218). Greenwich, CT: JAI.

National Housing Board. (1987). *Monograph on the human settlements situation and related trends and policies* (Ministry of the Environment). Helsinki: Government Printing Centre.

Statistical Office of Finland. (1986). *Statistical report KT 1986*. Helsinki: Author.

Statistical Yearbook of Finland. (1982-1987). *OSF XXXI, Municipal finances. Statistical reports JT (CSO)*. Helsinki: Government Printing Centre.

Weber, M. (1987). *General economic history*. New York: Greenberg. (Original work published 1927)

Villadsen, S. (1986). Local corporatism? The role of organisations and local movements in the local welfare state. *Policy and Politics, 14*(2), 247-266.

Ylönen, A. (1985). Societal factors shaping the internal structure of Finnish cities. In W. van Vliet, E. Huttman, & S. Fava (Eds.), *Housing needs and policy approaches: Trends in thirteen countries*. Durham, NC: Duke University Press.

6

Fiscal Retrenchment and the Relationship Between National Government and Local Administration in the Netherlands

ANTON KREUKELS
TEJO SPIT

Local Problems in the Netherlands[1]

As in most countries, the 714 municipalities in the Netherlands have been confronted with fiscal strain. For a thorough understanding of how municipalities have coped with fiscal stress, it is very important to know something about the way public management and public finance are organized in the Netherlands. One of the most striking features seems to be the dependence of Dutch municipalities on the national government for the implementation of their own policies. The extensive and intensive nature of relationships between Dutch municipalities and national government within a system of implementation of national policy at local/regional level by the municipalities is striking.

Although the diversity of both municipal activities and problems is enormous, the margin in which municipalities can implement their own policies is relatively small. Much research has been done about local problems, but most of this research has focused on only one problem. Recently, the Union of Dutch Municipalities (1986a) published a study concerning the cumulative aspects of local problems in the 17 largest municipalities in the Netherlands. In this report it is argued that an isolated approach to local problems is not a sufficient method for studying the overall effects of local problems. In the authors' view, local problems should be studied in relation to each other and the

TABLE 6.1

Ranking Scores on Social Problems in the 17 Largest Municipalities in the
Netherlands in Relation to Each Other

	(1)	(2)	(3)	(4)	(5)	(6)	(7)	(8)	(9)	(10)
Amsterdam	1		4	$\frac{1}{2}$		1	5	1	1	7
Rotterdam	5		2	$\frac{1}{2}$	$\frac{5}{6}$	3	2		5	7
The Hague	4		1	5					4	4
Utrecht	2	2		$\frac{5}{6}$		1	5			5
Eindhoven						3				1
Groningen		1		4	3			4		4
Tilburg		5			4					2
Haarlem	3		3						3	3
Nijmegen		4		3	1	2		3		5
Apeldoorn										
Enschede						$\frac{4}{5}$				1
Zaanstad										
Arnhem			5		2		4			3
Breda								2		1
Maastricht						$\frac{4}{5}$				1
Dordrecht										
Leiden		3							2	2

SOURCE: Union of Dutch Municipalities (1986a, p. 202).
NOTE: The absolute amounts in all the tables for this chapter are not controlled for inflation, because inflation in the Netherlands has been very low for the years concerned (between –1% and 3%, 1983-1987). Social problems: (1) decrease in population; (2) proportion of young people (15-24) in the population; (3) proportion of elderly people (65 and older) in the population; (4) proportion of people living at minimum existence level; (5) proportion of people looking for work; (6) proportion of people who benefit from social security; (7) proportion of people who work; (8) crime; (9) proportion of old housing stock; (10) number of times the municipality is in the top five.

emphasis should be on the cumulative aspects of these relations. Table 6.1 summarizes the ways in which local problems are cumulating in the 17 largest municipalities. The numbers in this table present a ranking score on each kind of problem. The report also argues that fiscal stress exacerbates these problems.

Rose and Page (1982) describe fiscal stress as a problem that arises when there are more claims upon the public purse than there is money to meet these claims (p. 1). According to this definition, fiscal stress can have many causes. In the Netherlands this appears to be the case in most municipalities. When asked what the major causes of fiscal stress in local government were, more than half—between 75% and 100%—of the local authorities considered national government policy to blame for their unhealthy fiscal situation. Furthermore, some 30% stated that national government policy was responsible to a large extent (50% to 75%) for their financial problems. The size of municipalities is an important factor in explaining these responses.

Larger municipalities are less dependent on national government policy than smaller ones, and are less likely to blame the national government. This is not surprising, as their policy margins are relatively larger than those of smaller municipalities.

When the Dutch municipalities were asked to name the major causes of their fiscal stress situations, the answers could be divided into two major categories. In the first category four factors are of major importance, three of which are directly connected to national government policy:

- reduction of resources of the Municipalities Fund
- reduction of funds for specific municipal situations, tasks, or projects
- implementation of a new law (in 1984) that regulates financial relations between municipalities
- increased costs of unemployment benefits

The fourth factor is partly due to national government policy and partly caused by an independent increase in unemployment. The increase in unemployment causes an increase in the municipal share of the unemployment benefits, which is not compensated by the national government.

The second category concerns municipal policy. Three factors are considered important:

- new projects that are important and cannot be postponed
- losses on land development
- losses from expired projects

The factors making up the first category are more important than those in the second.

The effects of fiscal stress must not be underestimated with relation to their indirect effects on other municipal problems. Due to lack of money for coping with difficult problems, municipalities are forced either to choose an often less-than-satisfactory solution or, worse, to neglect or minimize the problem. The increasing effect of fiscal stress on neighborhoods, where problems accumulate, is very marked.

The Policy Context in the Netherlands

Constitutional Framework and Public Finance

In understanding the Dutch municipalities' handling of fiscal stress, it is important to know something about the way public management is organized in the Netherlands. In the Netherlands there are three levels of public administration: national, provincial, and local. In this chapter we will concentrate on the lowest administrative level and try to describe and analyze how Dutch municipalities handle fiscal stress in their own political setting, with special emphasis on the relation between the national and local level of administration. As might be expected, this relationship has strong historical roots.

In the seventeenth and eighteenth centuries, the system of public administration in the Netherlands was very decentralized, with the provinces and municipalities in a strong position. The present-day constitutional system still provides real autonomy for provinces and municipalities within the limits laid down by the national government. The Dutch administrative system has even been characterized as a decentralized unitary state. However, this constitutional autonomy of the provinces and municipalities is limited by the precedence of national policymaking when confronted with provincial or local policymaking. Furthermore, the municipalities or provinces have to implement national programs and public services. Also, this involvement (*medebewind*) supersedes the domain of autonomy of the Dutch provinces and municipalities.

A second feature, dating back to the nineteenth century — the period in which the present-day Dutch constitutional system originated — is the uniformity of Dutch municipal administration. Thorbecke, the founding father of the nineteenth-century constitution, explicitly chose the principle of uniformity. Each municipality, be it a large city or a small village, is part of the same administrative system that circumscribes autonomy and uniformity. From the nineteenth century on, the autonomy of municipalities in general and that of local finances in particular became increasingly weaker. With regard to local finances, this began in 1865 with the abolishment of local excises as an independent source of local revenue (Hendrikx, 1981, p. 81). The last step on this path was taken in 1984, when the laws regulating the financial relations between central and local government (*Financiële Verhoudingswet*) were changed to reinforce an even more centrally guided financial system. In 1865, with the abolishment of local excises, the

TABLE 6.2

Change in Composition of Local Resources (in percentages)

	1932[a]	1953	1960	1970	1980
Municipalities Fund, discretionary grants included	25	53	34	38	32
Special purpose grants	25	30	58	55	62
Own resources and taxes	50	17	8	7	6
Total	100	100	100	100	100

SOURCE: Havermans (1985, p. 253).
a. Estimated.

Municipalities Fund (*Algemene Uitkering*) was introduced and, later on, grants for specific purposes (*Specifieke Uitkeringen*). The latter grants gradually became the major source of income for local governments in the Netherlands.

Municipal income in the Netherlands can be determined by adding up the following revenues:

(1) the Municipalities Fund

(2) the refinements (*Verfijningen*)

(3) the special purpose grants

(4) local taxes, fees, and user charges

As indicated by these categories, national revenues are supplemented by local taxes and fees and user charges that form additional sources of revenue for local government. Of all local taxes, the taxation on real estate is by far the most important: 69% of all own local sources are derived from taxes on real estate. In Table 6.3 the proportion of income from local taxes is shown for the four largest municipalities in 1984 and 1985.

As can be seen in Table 6.2, local revenues in the Netherlands continue to shift toward greater central control. Some argue that local taxation in the Netherlands should remain a relatively small share of local revenues in order to ensure a stable allocation of welfare over the entire nation (e.g., van der Dussen, 1975; Goedhart, 1975). However, other countries that share this concern (e.g., Denmark, Sweden, Norway) do not show such a pronounced pattern.

As can be seen in Table 6.4, local taxes, expressed as a percentage of total revenues, vary considerably. The percentage of local income in

TABLE 6.3

Income per Capita of the Four Largest Municipalities
from the Taxation on Real Estate (expressed in guilders; 1 guilder = $0.38)

	1984	1985	Percentage Change 1984-1985
Amsterdam	555.33	586.29	5.58
Rotterdam	774.76	720.12	−7.05
The Hague	419.39	460.67	9.85
Utrecht	555.15	584.83	5.35

SOURCE: Civil Service (1984, p. 3).

Dutch municipalities is extremely low when compared to the percentages of local income of municipalities in other European countries. The Netherlands is often regarded as a highly progressive country, closely resembling Sweden in terms of its organization as a welfare state. However, it appears — when compared to other progressive countries — to be surprisingly centralized as far as the fiscal relations between national and municipal government are concerned. As far as the national government is concerned, it could very well be characterized as patronizing. In this sense it is important to consider not only the extent to which municipalities depend on state revenues, but especially the conditions under which they are obtained. In the Netherlands these conditions are very strict.

The main local financial resources are grants from higher authorities. These grants can be divided into three groups, according to the new law of 1984 regulating the financial relations between the national government and the municipalities (*Financiële Verhoudingswet 1984*) (Koopmans, 1983, pp. 105-124). The first two categories together form the so-called Municipalities Fund (*Gemeentefonds*).

(1) The Municipalities Fund allocates funds on the basis of a number of characteristics of each municipality, including the following:

- a fixed amount, which is the same for every municipality
- a fixed amount for each acre of land or water, which can vary from year to year
- a fixed amount for every square meter of built-up area
- an equal amount for all municipalities, based on the number of inhabitants
- an equal amount for every house, whereby the number of houses are divided into categories (numbers) and the amount rises progressively with every category
- an amount for every house, according to the average height of all buildings

TABLE 6.4

Local and Other Sources of Local Government Income (in percentages)

	Local Taxes	Other Local Sources	Grants from Higher Authorities	Total
The Netherlands	5.5	0.5	94.5	100
Denmark	39.4	3.7	56.9	100
Italy	59.7	32.3	8.0	100
Norway	54.0	31.1	14.9	100
Sweden	45.8	29.1	25.1	100
United Kingdom	33.8	20.2	46.0	100
West Germany	34.3	36.1	29.6	100

SOURCE: Hoogerwerf (1980, p. 330).
NOTE: Other local sources are fees and user charges.

(2) Complementary to the Municipalities Fund are additional grants, called discretionary grants (*Verfijningen*). These grants are meant to make the Municipalities Fund more fitting to the specific needs of every individual municipality. They can also be divided into three major categories:

- Structural funding, which compensates for expenditures relating to structural elements in municipalities, such as poor soil or land marks.
- Functional funding, which is for the functioning of a municipality within its district (the function, such as education or health, can be disproportional to the size of a municipality, as in a municipality with a university within its territory). The most important distinction between this and structural grants is that functional grants can be influenced by municipal policy (e.g., by expansion of functions) and structural ones cannot.
- Friction funding, which is meant to help municipalities solve direct problems, such as a sudden decrease in population. These grants have a temporary character and are meant only to assist municipalities in special circumstances. In that respect they resemble the grants of the third category.

(3) Grants of the third category are special purpose grants or subsidies and are meant to finance specific municipal activities, such as road maintenance. The grants are issued by several departments of the national government. The number and the size of special purpose grants is considerable. In 1980 there were no fewer than 532 kinds of special purpose grants. Municipalities see this kind of grant as an ever increasing burden of centralization. These grants have been the major issue in

recent discussions about the financial relation between state government and local government. The Council of Municipal Finance estimated in 1981 that the share of these grants in the total of municipal budgets from 1970 to 1980 rose from 54% to 62% (see Table 6.2).

Until now we have discussed only local resources. Local expenditure — the other side of the balance sheet — will be presented in relation to municipal tasks. In doing so, we must keep in mind that larger municipalities have a more diversified set of tasks than the smaller ones. They therefore spend relatively more public money than the smaller municipalities.

Relation Between Organization of Public Finance and Municipal Activities

The relation between the organization of public finance in the Netherlands and the set of activities each municipality has to perform can be formulated as follows (Havermans, 1985, pp. 250-251):

(1) activities that the local administration has to implement and that are fully financed by the national government (e.g., social security)

(2) obligatory activities that are financed through a general fund (the Municipalities Fund)

(3) activities that municipalities are not obliged to implement but that are subsidized by the national government, mostly through special purpose grants

(4) activities that are not obligatory and for which the national government does not have a financial arrangement (paid for by local taxes or fees)

These four distinctive types of local activities, in combination with the way the finance is organized, influence how municipal activities are carried out and the potential for retrenchment at the local level. Theoretically, at least, these four distinctive types are important in local retrenchment policy. Municipal retrenchment policy will concentrate least, relatively speaking, on those activities supported by the central government. Furthermore, local activities will be better and more efficiently organized if local government is financially committed. Otherwise, if there are no financial implications for local government for the way in which the activities are organized, they are unlikely to consider efficient organization very important.

To illustrate the first statement: A municipality benefits most by cutting back on those activities that are chargeable to its own budget (e.g., category 4). The results of such an operation can immediately be

TABLE 6.5

Expenditures of Larger Dutch Municipalities, Divided by Category of
Activities and Size (expressed in guilders per capita; 1 guilder = $0.38)

Category	Population 400,000 and More	100,000 to 400,000	50,000 to 100,000	All Municipalities
General management	231.18	223.64	236.59	213.99
Public safety	422.85	223.00	212.69	165.32
Public health	118.77	82.58	58.63	63.12
Housing	373.09	277.84	361.36	282.13
Public works	596.37	278.94	269.43	316.39
Municipal properties	174.11	17.21	21.69	37.37
Education	1,235.08	1,001.36	481.67	422.07
Art and recreation	309.96	207.24	159.40	151.30
Social security	1,601.07	462.77	323.44	511.53
Economic affairs	265.04	56.42	28.72	51.78
Total[a]	5,153.30	2,540.48	2,423.91	2,465.65

SOURCE: Hoogerwerf (1980, p. 334).
a. Includes some activities that are not categorized.

seen in the budget. Other ways of cutting down expenses can be found
by carrying out obligatory activities more efficiently or by decreasing
the quality of these activities (categories 2 and 3). However, cutting
down expenses in this way is less effective in the short term. It will take
longer to see the positive results in the budgets. Besides, the size of
these cutbacks will have to be much larger to match the budgetary
impacts of cutbacks in local programs. It is evident that cutting back on
the first category will have at the most a marginal effect on the municipal budget. Even worse, as most activities in this category belong to
the obligatory services that each municipality has to provide by law, the
possibilities in this category are marginal.

Of course, retrenchment is also determined by the nature of the
activity and the size of the municipal budget for each activity. Table 6.5
reports how many guilders (per capita) Dutch municipalities spent per
category of activities in 1978. The major expenditure categories are,
respectively, social security, education, public works, housing, general
management, and public safety. On the whole, the largest municipalities spend more than twice as much money as the smaller ones, but the
largest municipalities have more functions to be financed within their

territories, especially in the fields of education, health, arts, and recreation. Except for general management, by far the greater part of the major expenditure categories are covered by special purpose grants from the national government; general management has to be financed out of the Municipalities Fund.

Finally, it should be noted that the national government has formulated some principles regarding the matter of centralization versus decentralization (Tweede Kamer, 1976-1977). The first of these is that there should be no separation of responsibilities among administrative levels in regard to the making of policy decisions and the way these decisions ought to be financed. But, as Tables 6.2 and 6.4 show, about 94% of municipal expenditure is not financed by local revenues.

The second principle is that if higher levels of government do finance lower government activities, these subsidies should come from the Municipalities Fund. Yet Table 6.1 shows that the trend is in the opposite direction: The proportion of financing by means of the Municipalities Fund is decreasing and the proportion by means of a great number of grants from separate departments of the national government is increasing.

The third principle states that on each administrative level there should be a satisfactory relation between the level of taxation and the level of services rendered by municipalities. Tables 6.2 and 6.4 again show that the town revenues from taxes, fees, and user charges add up to only 6% of municipal revenues.

The fourth principle emphasizes the degree of freedom each level of government should have in implementing its tasks in a responsible way. At the moment, however, about 60% of municipal revenues consists of special purpose grants, provided by the separate departments of the national government, with strict conditions attached. Even worse, national budget cutbacks concentrate far more on general funding than on the whole set of special purpose grants.

The fifth principle states that tasks and responsibilities should remain at lower administrative levels, unless it is demonstratively clear that they should be carried out by a higher administrative level. In reality there is a noticeable creeping tendency toward centralization, mainly visible in the way the national government monitors local government policy implementation.

The sixth and last principle is complementary to the fifth. It states that tasks and responsibilities that are now performed by higher authorities but that could as easily be performed at a lower administrative level should be turned over to that level. This rule, too, is in sharp contrast to reality.

The conclusion is clear. Although the wish to decentralize is manifest on all administrative levels, there is a growing discrepancy between this intention and ongoing developments.

Relationship Between National Government and Local Administration

The dependence of local administration on national government for financial resources has been the subject of the previous discussion. However, there are more ways in which the dominant role of the national government in public administration manifests itself. For example, in contacts between local and national officials, initiatives to establish person-to-person contacts almost always originate on the local level, with a direct purpose. Empirical studies indicate that contact between the national government and local authorities is often established through the mayor (e.g., Hoogerwerf, 1980, p. 339). Contacts between Dutch mayors and the Department of Housing were most frequent, with 75% of all Dutch mayors visiting this department at least once every three months. Furthermore, mayors of larger municipalities visited departments of the national government more often than did mayors of smaller municipalities.

Contact between levels of government is not always direct: It also occurs through political parties. As far as the larger parties are concerned, they are the same at the local as at the national level, represented in parliament as well as in city councils. However, most important in the relationship between the two levels of government is the Union of Dutch Municipalities. All 714 Dutch municipalities are members of this union, which was founded in 1912 as the result of a discussion between some mayors waiting in the anteroom of a government department. The union could be described as the platform of Dutch municipalities to influence national government policy in order to promote the functioning of local public administration. No empirical studies have yet been made as to the efficacy of the work of the Union of Dutch Municipalities, but former Prime Minister Van Agt described it as "one of the most powerful pressure groups in the country." This position was recently consolidated with the Administrations Agreement (*Bestuursakkoord*) between the Union of Dutch Municipalities and the State Department. One of the arrangements in this agreement is that in the next four years, a considerable amount of money (about 50 million guilders) will be added annually to the Municipalities Fund. This increase is meant to be spent on capital works and unemployment projects. In addition to this pressure group function, the Union of Dutch Municipalities also has a service function for its members. The service function consists of the provision of individual advisers and specialized technical support in different policy areas for its members, the Dutch municipalities.

A very important advisory body in the relation between national and local government is the Council of Municipal Finances — instituted by law in 1960 — the task of which is to advise the several departments of the national government in matters concerning municipal financial affairs. The chair of this council is appointed by the Union of Dutch Municipalities, and the appointment must be approved by the State Department and the Department of Finance. The secretariat is situated in the office of the Union of Dutch Municipalities. The influence of this council varies. Each department of the national government reacts independently to the council's recommendations; for instance, the influence can be very clearly felt in the field of arts and recreation in contrast to the field of education, where it is practically zero (Hooger-werf, 1980, p. 341). Alongside this task, the council supervises the financial division of the Municipalities Fund among the municipalities.

Municipal Retrenchment Policy in the Netherlands

The Magnitude of Local Fiscal Retrenchment

In recent years retrenchment has been the subject of many discussions at various levels of government. In these discussions Dutch municipalities have always fought against their financial dependence on the national government. Furthermore, they have argued strongly against the measures — issued by law — of the national government requiring them to reduce their expenses. On this point Dutch municipalities were supported by the Council of Municipal Finances, which pointed out that the share that municipalities contributed to the overall government retrenchment efforts was totally unfair and out of proportion. Besides, the extent to which municipalities in general had to economize was disproportionately greater than that of the national government. Of course, not every municipality suffered under the same burden of retrenchment. Some encountered severe financial problems, while others were apparently only slightly troubled by national retrenchment policy. From the perspective of the financial relations between the national government and local administration, it is clear that national government retrenchment policy is the main cause of local budget problems. As such, there is very little choice for Dutch municipalities.

To determine fiscal stress and the various retrenchment strategies adopted by Dutch municipalities, we surveyed all Dutch municipalities in 1985. In total, 294 of the 441 municipalities responding to our questionnaire claimed that they had decreased their expenditures in the three-year period (1983-1985) we investigated: 191 (43%) cut expenditure every year, 61 (14%) did so only two of the three years, and 42 (9.5%) made cuts only once during this period. These last 42 municipalities were mainly very small, with fewer than 10,000 inhabitants each.

A distinction must be made between structural and incidental retrenchment measures. Structural measures are defined as measures for which, every year, a certain amount of money is withheld (such as stopping certain activities), while incidental measures are interpreted as measures from which a municipality benefits for only one year. Furthermore, it seems logical that smaller municipalities can more easily make ends meet with incidental retrenchment measures. In contrast, the larger municipalities almost always have to turn to structural retrenchment measures for more than one year. In general, structural retrenchment measures—which cut down municipal expenditure for many years—obviously carry much more weight than incidental retrenchment measures. This applies both to the number of municipalities that take structural retrenchment measures and to the absolute impact of such measures. The absolute amounts for both types of retrenchment measures are shown in Table 6.6.

Assuming that the municipalities in our research sample constitute a fair representation of all Dutch municipalities, it is possible to estimate the total amount of money involved when municipalities in the Netherlands cut back on their budgets during these three years. In this estimation structural cutbacks are especially important because this type of retrenchment recurs every year in the municipal budgets. Based on the figures shown in Table 6.6, Dutch municipalities cut their budgets effectively by 600 million guilders in 1983, by 825 million guilders in 1984, and by 877 million gilders in 1985. In three years' time, Dutch municipalities spent 4.3 billion guilders less than they would have done without their obligation to decrease expenditure.

TABLE 6.6

Incidental and Structural Retrenchment[a] in 1983, 1984, and 1985

(expressed in guilders; 1 guilder = \$0.38)

	1983	1984	1985
Incidental retrenchment	29,024	34,974	35,397
(N)	(53)	(61)	(54)
Structural retrenchment	359,459	495,069	526,510
(N)	(207)	(263)	(229)
Total amount	388,483	530,043	561,907
Total number	(222)	(263)	(246)[b]
of municipalities			

a. A distinction must be made between structural and incidental retrenchment measures. Structural measures are defined as measures for which, every year, a certain amount of money is withheld (such as stopping certain activities), while incidental measures are interpreted as measures from which a municipality benefits for only one year.
b. It should be noted that there are also municipalities that take both kinds of retrenchment measures.

Municipal Strategies: A General Impression

In theory, municipalities have available many strategies for coping with retrenchment. In our research — which, in general, is representative for all 714 Dutch municipalities — we chose 17 of the most likely strategies and asked the chief administrator officer (CAO) in each municipality to assess their importance. In the Netherlands the CAO is not elected, but is appointed by the city council. He or she is the highest official in local government and occupies a key position between the other officials, the mayor and alderman, and the city council. In light of this fact, he or she is in a more or less neutral position. This neutrality, and the overview of municipal finance, makes the CAO the most reliable source of information.

Table 6.7 reports the relative importance of each of the 17 strategies according to the chief administrators' rating of each strategy as one of the most important (4), very important (3), rather important (2), or not at all important (1).

When the average score is set at 2.00 over these three years, some strategies appear to be more important in municipal retrenchment policy. (The scores in this table imply that the strategies are actually used.) The table gives the average score for these strategies for each year separately and for the three years together.

The choice between strategies can be influenced by many factors. The most important ones are the size of municipalities and their politi-

TABLE 6.7

Average Scores of Dutch Municipalities on Categories of Retrenchment Policy
in 1983, 1984, 1985, and Average Score over All Three Years

Strategy	Average Score			
	1983	*1984*	*1985*	*All Three Years*
Revenue				
Seek new nonlocal revenues	1.40	1.48	1.46	1.45
Raise local taxes	2.24	2.12	1.93	2.10
Increase fees and user charges	2.23	2.25	2.13	2.20
Draw down surpluses	2.13	2.28	2.36	2.26
Expenditure				
Make cuts in all policy areas	2.64	2.72	2.55	2.65
Improve efficiency	2.67	2.55	2.57	2.59
Reduce working hours	1.22	1.29	1.36	1.29
Encourage early retirement	2.33	2.59	2.46	2.46
Lay off personnel	1.11	1.15	1.13	1.13
Use volunteers	1.49	1.56	1.56	1.54
Shift responsibilities to other units of government	1.09	1.11	1.15	1.12
Contract out services	1.55	1.77	1.80	1.71
Decrease the quality of services	1.67	1.79	1.85	1.77
Eliminate services	1.40	1.48	1.51	1.46
Reduce control of regulations	1.33	1.38	1.39	1.37
Reduce capital expenditure	2.31	2.44	2.46	2.40
Postpone payments	1.17	1.17	1.18	1.17
Other	2.19	2.45	2.29	2.31

NOTE: This table reports the relative importance of each of these strategies according to the chief administrators' rating of each strategy as one of the most important (4), very important (3), rather important (2), or not at all important (1). When the average score is set at 2.00 over these three years, some strategies appear to be more important in municipal retrenchment policy. The scores in this table imply that the strategies are actually used.

cal composition. The size of municipalities is expressed in the number of inhabitants; the political composition is expressed in percentages of Social Democratic council members. Tables 6.8 and 6.9 illustrate the most important strategy in relation to size and political composition. For the other strategies only the interpretation will be given.

(1) *Cuts in all policy areas* (2.65): The figures show that this strategy was very important, especially in the larger municipalities. Also, this strategy is more often adopted by left-wing local government than center or right-wing politically oriented local governments.

TABLE 6.8

Number of Municipalities Using Strategy "Cuts in All Policy Areas"
in Relation to the Size of Municipalities, 1983-1985

Number of Inhabitants	Most Important	Very Important	Somewhat Important	Least Important	Don't Know	Total
0-10,000	12	20	39	57	71	199
	(6.03)	(10.05)	(19.60)	(28.64)	(35.68)	(100)
10,000-20,000	10	29	28	27	19	113
	(8.85)	(25.66)	(24.78)	(23.89)	(16.81)	(100)
20,000-50,000	7	32	26	14	6	85
	(8.24)	(37.65)	(30.59)	(16.47)	(7.06)	(100)
50,000-100,000	0	7	5	4	0	16
	—	(43.75)	(31.25)	(25.00)	—	(100)
100,000 and more	2	7	1	1	0	11
	(18.18)	(63.64)	(9.09)	(9.09)	(9.09)	(100)
Total	31	95	99	103	96	424
	(7.31)	(22.41)	(23.35)	(24.29)	(22.64)	(100)

NOTE: Percentages appear in parentheses.

(2) *Efficiency improvement* (2.59): This strategy proves very important in larger municipalities. As could be expected, this strategy is less important in smaller ones. The political composition is not very relevant to the importance of this strategy.

(3) *Early retirements* (2.46): This strategy is of no importance in municipalities with fewer than 10,000 inhabitants. The larger a municipality, the more important this strategy becomes. Again, differences in political composition seem unimportant.

(4) *Reduction of capital expenditure* (2.40): This strategy is widely used by all municipalities; neither size nor political composition is important.

(5) *Drawing down surpluses* (2.26): Municipalities with more than 50,000 inhabitants consider this strategy far more important than the smaller municipalities. Differences in political composition are again of no importance.

(6) *Increases in fees and user charges* (2.20): This strategy shows almost the same pattern as drawing down surpluses, except that the difference between larger and smaller municipalities is much more diffuse. The political composition shows little effect on the adoption of this strategy.

(7) *Increases in local taxes* (2.10): In general, this strategy was of equal importance in municipalities of all sizes. In municipalities with a right-wing local government, however, this strategy was slightly more important.

TABLE 6.9

Number of Municipalities Using Strategy "Cuts in All Policy Areas"
in Relation to the Political Composition of Municipalities, 1983-1985

Political Composition	Strategy Ranking					
	Most Important	Very Important	Somewhat Important	Least Important	Don't Know	Total
0-20% left-wing	14	35	39	54	52	196
council members	(7.14)	(18.88)	(19.90)	(27.55)	(26.53)	(100)
20-40% left-wing	14	37	44	35	36	166
council members	(8.43)	(22.29)	(26.51)	(21.08)	(21.69)	(100)
More than 40%	5	26	19	15	13	78
left-wing council members	(6.41)	(33.33)	(24.36)	(19.23)	(16.67)	(100)
Total	33	100	102	104	101	440
	(7.50)	(22.73)	(23.18)	(23.64)	(22.95)	(100)

NOTE: Percentages appear in parentheses.

As for the category "other retrenchment measures" (2.31), two additional strategies were often mentioned, namely, stopping new projects and selling municipal property.

As suggested above, the way municipal activities are financed goes a long way in explaining the way in which Dutch municipalities have implemented retrenchment policy. A random sample of 60 municipalities were asked the net amounts by which they had cut back on ten different sectors of urban policy. These figures are shown in Table 6.10. The last category in this table consists of all retrenchment measures that do not fit into the other sectors. The organization of municipalities in the Netherlands is very diverse. Considering this diversity, it is not so surprising that the score in this category is high. The arts and recreation category was subject to most of the urban retrenchment measures in 1983, but in the years following, retrenchment measures had a more proportionate effect on this sector. Apparently there is not much left to economize on.

Discussing retrenchment in sectors of policy does not say anything about the way in which these retrenchment measures were implemented. From the Bank of Dutch Municipalities, signals were sent out that municipal capital works were sharply decreasing in all areas of municipal policy. Although reduction of capital expenditure was already indicated as an important strategy, the total impact of this strategy might be far more important than at first estimated. Therefore, the development of municipal capital works was examined in the 60

TABLE 6.10

Measures Through Which Urban Retrenchment Has Affected
Main Urban Policy Areas (expressed in thousands of guilders; 1 guilder = $0.38)

Sector of Urban Policy	Average 1983	Average 1984	Average 1985	Total
General management	442.8 (23.29)	836.0 (44)	622.1 (32.71)	1901.7 (100)
Public safety	154.7 (30.52)	187.2 (36.94)	164.9 (32.54)	506.8 (100)
Public transport	591.5 (35.44)	590.9 (35.40)	486.8 (29.16)	1669.2 (100)
Public health	458.7 (40.47)	325.7 (28.74)	349.0 (30.79)	1133.4 (100)
Housing	561.9 (42.33)	409.1 (30.82)	401.4 (30.24)	1327.4 (100)
Education	542.0 (36.88)	517.2 (35.20)	410.3 (27.92)	1469.5 (100)
Art and recreation	971.7 (46.09)	758.5 (35.99)	378.0 (17.93)	2108.2 (100)
Social security	549.0 (38.08)	410.6 (28.48)	482.0 (33.44)	1441.6 (100)
Economic affairs	593.2 (47.94)	289.7 (23.41)	354.4 (28.64)	1237.3 (100)
Financial affairs	533.2 (34.72)	371.8 (24.21)	630.9 (41.08)	1535.9 (100)
Other	1102.2 (50.98)	572.8 (26.50)	486.5 (22.51)	2161.5 (100)

NOTE: Percentages appear in parentheses.

municipalities during a three-year period (1982-1984). The figures are shown in Table 6.11.

From these figures one can easily see the sharp decrease in municipal capital works in the Netherlands. In this three-year period Dutch municipal capital works decreased by almost a third (29.3%). The seriousness of this situation — postponing problems until a future period — has been recognized by both local and national administrations and has already had effects (see, e.g., the Administration Agreement).

What major conclusions can be drawn from these findings? A first conclusion must be that differences in political composition are apparently relatively unimportant in responding to fiscal stress. Second, the importance of the size of municipalities is somewhat more relevant,

TABLE 6.11

Decline of Capital Works in Dutch Municipalities in the Period 1982-1984
(expressed in millions of guilders; 1 guilder = $0.38)

	Number of Municipalities	Net Amounts Spent on Capital Works	Average
1982	60	953.75	158.9
1983	60	850.42	141.7
1984	60	674.11	112.3

but does not influence choices among strategies a great deal. The ways of coping with fiscal stress are not very diverse. A possible explanation for this lack of diversity lies in the very small margins left to municipalities with regard to choosing and implementing their own retrenchment policies.

Municipal Strategies: A Closer Look

The margins within which Dutch municipalities can set their own independent policies are relatively small because the financing of most activities comes from the national government. This probably accounts for the fact that the diversity in strategies among Dutch municipalities was not very large. As mentioned above, most municipalities chose, from the 17 strategies, a mixture of measures. Individual circumstances determined which strategy measures were emphasized. The policy sector in which the expenditure strategies (the most important ones) were used, however, varied as described in Table 6.9.

In order to obtain a better understanding of the way in which these retrenchment measures have affected Dutch society, four case studies were conducted (Union of Dutch Municipalities, 1986b). Each case study focused on a different sector of municipal policy:

- public works, including maintenance of roads and squares
- maintenance of public parks
- art and recreation
- social security and social welfare policy

For each case study four municipalities were investigated.

Briefly, the case study on public works policies showed that these four municipalities had been economizing on the maintenance of public roads by means of a decrease in the standard of maintenance. In addition, three of these four municipalities had reorganized their public works sections, which led to a reduction in the number of employees. The real effects of these policies are not yet physically visible, but one can expect a decrease in safety on the roads in the future, higher maintenance costs, and probably an increase in complaints and claims from citizens.

In the case study on the maintenance of public parks, the immediate effects are more directly visible. The organization of departments responsible for the maintenance of public parks has been the target of retrenchment measures. Strategies to improve efficiency and to contract out parts of the maintenance activities were the main tools used. Apart from this, the standard of maintenance also has decreased. These case studies signal the fact that retrenchment in public works is likely to lead to a decline in employment in the short term and, in the long run, a cumulation of negative effects as a result of the combined retrenchment measures.

Municipal activities related to the field of culture and arts can vary from education to museums. Four stages of local response could be distinguished in the case studies: (1) a search for new income sources and the use of local reserve funds; (2) economizing on personnel, in terms of both number of employees and salary levels; (3) decrease in the quality of the product, without the continuation of the activity being questioned; and (4) evaluation, with the result that certain activities are terminated or drastically reorganized. Although these stages can be observed in each of the investigated municipalities, it seems clear that larger municipalities are making more use of the later stages than are smaller ones. The citizen directly experiences retrenchment effects only in the third stage. Long-term effects are unpredictable. Some even claim that the future might bring some sort of *Verelendung*, which is a Marxist concept that sometimes the conditions have to be very bad for the creation of something extraordinary.

For Dutch municipalities, social welfare activities primarily involve the implementation of national social security legislation. As a result of the economic recession, municipal activities in this field have been increasingly expanding. The numbers of citizens appealing to municipalities have increased, and the nature of their problems has worsened. In general, municipalities have tried to avoid the cutback of social welfare in their retrenchment strategies. Most of the municipalities examined had even expanded this sector; however, this expansion was disproportional when compared to the increase in work load for local social security officials. The quality of welfare has been severely damaged by the enormous work load in combination with the retrenchment measures. The four municipalities examined also complained about the complexity of, and sudden changes in, procedures and social security measures imposed in national legislation by the national government. This will be even more critical in the future if reorganization, necessary to cope with these complexities and dynamics amid financial cutbacks, is not realized.

Explaining Policy Innovation

An Initial Assessment of Fiscal Austerity Policy

There is, in the Netherlands, little research on fiscal austerity and policy innovation at the local level. The survey we developed and the findings reported here are the only broad inventory currently available. In this concluding section, we give a more qualitative assessment, first of fiscal austerity policy and subsequently of policy innovation and the political implications.

First we must note that variations among municipalities with regard to retrenchment strategies are, in general, small. We have already suggested one possible explanation — the pronounced centralization in the Netherlands, which offers little opportunity for discretion at the local level.

Nevertheless, one main division among the municipalities is evident in the data: those municipalities with a certain amount of surplus and those without such a financial buffer. The differences with regard to fiscal austerity policy between these two categories are marked. In general, those in the first category got through in the period of retrenchment (1983 to the present) following the economic recession (1980-1983) even when they had to adapt to a regime of fiscal austerity. Within the second category one can distinguish two classes: (1) municipalities that were able to adapt to the new situation by means of a package of different measures, related first to expenditure and second to revenues; and (2) municipalities that also had to reduce the number of municipal employees dramatically. In general, the medium and large cities fall into the second category, while the small municipalities are overrepresented in the category with a firm financial buffer. The division between larger and smaller municipalities, noted several times, therefore reflects this distinction.

Political composition at the local level plays only a limited role in retrenchment policies. Even if some right-wing and center regimes were more prone to raise local taxes, in general, the variation between right- and left-wing regimes at the local level is really very small. How can that be explained? At the national level, surprisingly, the contrast between the governing coalition of right-wing and center political parties and the opposition of left-wing parties is great with regard to fiscal austerity policy. While even the left-wing parties at the national level feel that the economic recession requires retrenchment measures and cutbacks in the public sector budgets, they are nevertheless still champions of a strong public sector. The same political parties are active at the local level as at the national level, but at the local level the dividing line between political parties with regard to fiscal austerity and retrenchment strategies is not as sharp as at the national level. In general, all political parties — be they right-wing, center, or left-wing — agree at the local level on the need to limit the consequences of the economic recession for citizens. In particular, one can observe in all political parties at the local level an active involvement in preventing a deterioration of the situation for the lower-income groups requiring social security payments. One must conclude that, at the moment, the Dutch local political machine functions as a shield for citizens in general and for the socially and economically dependent in particular; this in a period when the policy of the national governing coalition is, at the same time, fiscally conservative and socially more severe.

From Fiscal Austerity to Policy Innovation

The following observation concerns the changing character of local fiscal austerity policies. As indicated before, Dutch local administrators clearly preferred the general retrenchment strategy of postponing activities and reducing investments. However, after an initial period (1980-1986) in which the attainment of a sound financial basis was the first priority, one can again observe a shift to future-oriented investments. This reorientation is related to a new climate of entrepreneurial politics in Dutch local administration that became manifest after 1983. The fiscal austerity policy on the local level is thus characterized by a gradual evolution of (1) cutbacks in specific policy areas (particularly the least vital ones, such as culture, arts, recreation), (2) more generic retrenchment strategies (higher efficiency, early retirement, and reduction in working hours of personnel, contracting out services and so on), and (3) creative financing (especially investments by means of public-private partnership, loans, and leasing). The present-day transition to creative financing marks an expanding process from fiscal austerity to policy innovation. Given the limited options on the revenue side due to the Dutch centralized administrative and financial system, this also means a broadening of the usual management measures (reorganization and so on) by means of other sources of finance, particularly by contracting with the private sector. The first retrenchment strategies and adaptations in the period 1980-1986 can be interpreted as catalysts for the most recent period of entrepreneurial policymaking at the local level.

This entrepreneurial policymaking is still of a very experimental nature, but the transition seems definitive. The larger and middle-sized cities function here as pioneers. Beneath the surface of new initiatives and projects one detects the gradual restructuring of the local political machine. It means a shift to a more mixed public-private system and a reemphasis on economic, technological, and cultural innovation, while preserving the system of social security and social policy, albeit in an adapted and reduced form. In this shift to a more flexible, less complex, and less extended local public sector, the actual agenda and conditions seem to be very important. Reference can be made here to the new technologies, the importance of international economic relationships, the impact of social and cultural segmentation at local and regional levels, demographic changes (e.g., an aging population), and relatively large and stable unemployment.

The present-day debate in the Netherlands mirrors this intention to find a new balance between the achievements of the Dutch welfare policy system and the new exigencies of economic, technological, and cultural innovation. There is a conviction that, in general, the public sector must become smaller. However, with regard to the possibility of greater local independence and less administrative and financial centralization, there is still a serious gap between the present-day attitudes of the local political machine and those of the national government, including the political parties, as far as the latter operate at the national level. While the national level is very reticent to give more autonomy to Dutch local administrations, local politicians and administrators are creatively seeking ways to enlarge their opportunities and options. This is currently occurring by means of new relationships with the private sector at the local level and by administrative contracts with the national government.

Political Effects in the Netherlands

Strategic Autonomy

With regard to the autonomy of Dutch municipalities, the 1980s have been a period of transition. The new balance between national government and local administration evolving at the moment means a departure from the traditional centralized system. The experiences with fiscal austerity policy indicate that the new ratio of freedom or oversight of the national versus the local administration is to a very large degree a function of the representation of the separate groups and interests at the local, provincial, and national levels. For this reason, we shall first make some observations about the changing balances among these groups and sections in relation to the Dutch administrative system, before presenting some conclusions about the relative autonomy between national and local administration.

One of the driving forces behind the centralization of the Dutch administration is the system of denominational segregation, characteristic of the Netherlands from the nineteenth century onward. This segregation was decisive for the different interests and sections of Dutch society. For instance, the labor unions were organized according to the most important denominations: catholic, reformed, rereformed in combination with the socialist and communist unions. This institutional arrangement meant a pronounced administrative and organizational segmentation, especially at the local and provincial levels. However, in the first half of this century the elites and top levels of the different denomination-based organizations and associations gradually began to coordinate their interests within the framework of the national government. This arrangement was referred to as the "pacification policy." The agreement on the freedom to offer education in accordance with the specific denomination became particularly famous. From the beginning of this century onward, an ever increasing network of interest groups organized itself at the national level and was firmly related to this pacification arrangement. From the 1960s onward, this denominational segregation diminished; however, the centralized national junction of interests and organizations was by then a cornerstone of the Dutch administrative system. The development of welfare policy, guided by the undercurrent of distributive justice for all Dutch citizens and organizations, reinforced the centralized administrative and financial system in the 1960s and 1970s.

The economic recession (1980-1983) and the necessary adaptation in the 1980s led for the first time to rearrangements in this traditional Dutch political system. Fiscal austerity and the reduction of the public sector did not cause these rearrangements, but supported them. At the moment one can observe the first public-private partnerships at the local level. These affect the traditional balance of a financially and administratively strong national government and a very dependent local administration. Once the private sector again begins to play an important role in relation to local administration, the dependence on the national government diminishes to some degree. Also, the shift in negotiations between labor unions and corporations from the national to a local and regional level is a supportive factor. At this point it is important to notice that in the Netherlands negotiations about wages of municipal employees is not a local concern, but a concern of the central government. At this level negotiations take place collectively for all civil servants. The same applies to initial efforts at administrative decentralization, such as urban renewal block grants. At the moment it is uncertain what the final outcome will be. One must not forget that the national government ultimately controls the other administrative levels and, indirectly, through its centralized taxes and juridical impact as a unitary state, the whole Dutch private sector. In the meantime one cannot deny the changes in the strategic autonomy of groups, officials, and administrators, relative to national and local levels.

Another important feature of the present-day restructuring is the stronger position — in terms of strategic autonomy — of the business sector, relative to the social services and social security sectors. Within the administrative system, the business sector is in a favored position at the moment. Its strategic autonomy is now greater than that of organizations and associations in the sector of social policy. The extent of assistance to the business sector through public resources is impressive. One can observe a whole range of experiments and tryouts here, such as the establishment of economic development corporations. Nevertheless, the limitations of current tax policy and administrative rules with regard to a more creative application of these financial "incentives" (although with special attention to public interest), are striking. Thus a large amount of Dutch capital flows like a "money drain" to other countries, especially to the United States, due to the limited opportunities for venture capital and real estate financing in the Netherlands.

A final observation concerns the necessary increase in knowledge and expertise with regard to finances and management in the local administration. Municipal authorities currently use public-private partnerships to obtain the necessary know-how. Subsequent projects benefit from this spin-off from public-private partnerships. These first experiments, however, show the seriousness of deficiencies in local administration with regard to market-related information.

Relative Autonomy

The discussion in the Netherlands about the reinforcement of local autonomy is very animated at the moment. Yet, as mentioned before, the structural conditions for strong local politics in the Dutch administrative system are not favorable. Our central proposition, however, is that the rearrangement of organizations and associations, relative to national and local administrations, results indirectly in a reinforcement of stronger local politics in the long term.

In the meantime, it is useful to refer to recent proposals for changes in the financial system of the municipalities, originating from the Council of Municipal Finances (1986):

(1) reinforcement of and more reliable availability, for the municipalities, of the Municipalities Fund, with, in the long term, a fixed revenue share relative to the national budget and, in the short term, a proportional share of national cutbacks instead of extra cuts

(2) radical reduction in the large number of special purpose grants tied to the separate departments of the national government in order to limit the extreme conditioning of local politics associated with these grants

(3) expansion of local taxes

In regard to the last, the Council for Municipal Finances is cautious, but responds to wishes in municipal circles to strengthen local taxes in order to promote more satisfactory allocations on the expenditure side (one assumes in the case of local taxes a more direct relation between revenues and expenditure and consequently more conscious choices with regard to expenditure). There is a preference for tax sharing in the unitary Dutch tax system, so the Council here suggests surcharges for local administrations on national income tax and national wage tax.

The Council for Municipal Finances is trying, by means of these proposals, to restore a more autonomous financial regime. This has gradually been declining since 1851, the year of the first law relating to present-day local administration. The adaptation of the financial system is one important element. Another prerequisite is a new attitude in which a bureaucratic approach is replaced by a style of "innovative financing." But as yet there is little progress within municipal circles as far as attention to investments (investments versus current accounts), more advanced forms of accounting and financing, leasing, depreciation, costs related to the management of public services, and so on are concerned.

Conclusions

To conclude, we look at the recent differentiation between policy areas with regard to relative autonomy between the national and local levels. Even when there is only a marginal adaptation in favor of local administrations, some policy sectors are more progressive than others. The most centralized policy sector at the moment is still that of education. Municipalities are reduced here to sheer agents of national policy. To a lesser degree, the same applies to the health sector. However, recent discussions have centered on the possibility of restoring relationships between the market and the Dutch health system so that tuning of supply and demand again becomes possible. This means a greater emphasis on the local and regional setting of health organizations, such as North American health maintenance organizations. The policy area of social security is also still very centralized. The Dutch social security system is, internationally speaking, one of the most centralized; here too the municipalities function as agents of national policy. However, particularly in the large cities, the pronounced accumulation of problems is leading to mutual dependence of the local and national administrations. Informal contacts between administrations already mirror the increasing role of the municipalities in the social security sector.

Two policy areas in the Dutch administrative system can be considered as laboratories of decentralization: urban renewal and housing, especially public housing. In both fields a financing and programming system has been designed in which local administration can make its own choices within a broad allocation plan. However, even in these two fields the strategy choices are still made at the national level. So, on the whole, the Dutch administrative system is still very hesitant about allowing local autonomy. That means that the examples mentioned above in general do not represent strategic choice or control on a local or regional level with regard to finance or personnel. However, there is more freedom than before with regard to the distribution of facilities. More significant changes can be expected from the changing relationships of groups and officials both with each other and with the national and local administration. As stated previously, these changes in strategic autonomy seem to have a decisive impact in the long term in favor of greater local autonomy.

Note

1. The Dutch FAUI research consisted of three stages. In the first stage, a survey was performed in 1985 among all 714 Dutch municipalities. Its aim was to obtain a general impression of the way Dutch municipalities were handling fiscal stress. A total of 441 questionnaires were returned, a fair representation of all Dutch municipalities relating to size and political structure. In the second stage, in the same year, a random sample of 60 municipalities was taken, in order to obtain a better understanding of the effects of fiscal retrenchment. In the third stage, 16 case studies were made. In order to do so, four policy areas were selected where retrenchment effects were expected to be most severe. For each policy field, 4 municipalities were selected, differing in size, toward which the case studies were oriented. The case studies took place in the first part of 1986. Most of the results of the research have also been published in Dutch, by the Union of Dutch Municipalities.

References

Civil Service (Binnenlands Bestuur). (1984). *No. 14*. The Hague: Author.

Council of Municipal Finances (Raad voor de Gemeentefinanciën). (1986). *Na een collier de misère: gelijkwaardigheid, een zilveren raad over gouden koorden.* [After a "collier de misère": Equality, silver advice and golden strings]. The Hague: Union of Dutch Municipalities.

Dussen, J. W. van der. (1975). *De allocatie van middelen en de financiële verhoudingswet* [The allocation of means and the law which regulates the relationship between national and local government]. Den Haag.

Goedhart, C. (1975). *Hoofdlijnen voor de leer der openbare financiën* [Guidelines for public finance]. Leiden: Stenfert Kroeze.

Havermans, A.J.E. (1985). Gemeentefinanciën [Municipal finance]. In W. Derksen & A.F.A. Korstein (Eds.), *Lokaal bestuur in Nederland* [Local government in the Netherlands]. Alphen a/d Rijn.

Hendrikx, J.A.M. (1981). [Chapter]. In N.C.M. van Niekerk (Ed.), *Macht en middelen in de verhouding rijk-lagere overheden* [Power and means in the relationships between national and local administration]. Den Haag.

Hoogerwerf, A. (1980). Relaties tussen centrale en lokale overheden in Nederland [Relations between the national government and local administration in the Netherlands]. In *Beleid en Maatschappij* [Policy and society]. Meppel: Boom.

Koopmans, L. en A. W. (1983). *Overheidsfinanciën* [Government finance]. Leiden.

Ministry of Finance. (1987). *The financial relationship between central and local government*. The Hague: Author.

Rose, R., & Page, E. (Eds.). (1982). *Fiscal stress in cities*. London: Cambridge University Press.

Tweede Kamer. (1976-1977). *Wetsontwerp reorganisatie binnenlands bestuur* (14322, No. 3) [Bill regarding the reorganization of civil service]. Memorie van Toelichting.

Union of Dutch Municipalities. (1986a). *Gemeentelijke heroverwegingen* [Municipal retrenchment policy]. The Hague: Author.

Union of Dutch Municipalities. (1986b) *Gevolgen van gemeentelijke ombuigingen* (IOO and VNG/SGBO) [Effects of municipal retrenchment policy]. The Hague: Author.

7

French Local Policy Change
in a Period of Austerity:
A Silent Revolution

VINCENT HOFFMANN-MARTINOT
JEAN-YVES NEVERS

In France, as in most European countries, the mid-1970s marked a turning point in the evolution of local government.[1] Change was not, however, expressed in situations of acute fiscal crisis, as was the case with New York and other American cities, nor by movements of protest against taxation or very aggressive political tensions as in Great Britain. French municipalities gradually adapted to the new climate brought about by the recession of 1974. A "quiet revolution" at the local level has developed, challenging the rather stereotyped image of French society and government inherited from Tocqueville. Over the past ten years, local officials have gained more autonomy vis-à-vis central government, but they have to run the communes in a more complex local context and in a climate where it has become more difficult to mobilize new resources in order to increase or even maintain services provided to citizens.

New Challenges to Local Government

French communes have not had to face the same sort of fiscal crisis as some American or British towns. Compared to the postwar period of economic growth, a certain easing off of the "local fiscal crisis" can be seen even since the mid-1970s. This crisis has been the subject of almost permanent polemic between local authorities and central government since the start of the century (Hoffmann-Martinot & Nevers, 1985).

1975-1983: Moderate Fiscal Stress

There are several factors that may account for the moderate nature of fiscal stress between 1975 and 1983:

* Central government policies during the presidencies of Valéry Giscard d'Estaing and François Mitterrand brought about alternating phases of economic boosts and austerity but did not lead to any deliberate policy of tax and public spending reductions as was the case in West Germany, Great Britain, and the United States. Between 1975 and 1983 taxes and social security contributions increased from 37% to 45% of the gross domestic product in France, from 36% to 37% in West Germany, and from 36% to 38% in Great Britain, and decreased from 30% to 29% in the United States. The increase was particularly marked during the term of office of the conservative President Giscard d'Estaing.
* France is one of the very few industrialized countries where transfers from central government to local governments have not noticeably decreased. For French local administrations on the whole, the share of resources received as a result of transfers increased from 39% in 1979 to 44% in 1982, whereas in the United States, West Germany, and Great Britain, it decreased (Wolman & Goldsmith, 1985). Table 7.2, concerning the budgets of communes alone attests to this fact.
* In France the proportion of local spending remains lower than in most other countries; total local expenditures account for approximately 8% of the GDP and 16% of public expenditures (against 19% of GDP and 37% of public expenditures in the United States, 13% and 26% in Great Britain, and 35% and 88% in Denmark). As regards local expenditures, commune expenditures account for approximately 50% and that of urban communes of more than 10,000 inhabitants 25%. In such circumstances control of local expenditures is less crucial, politically and economically, than in other countries to the success of a policy of austerity and reduction of taxation.
* Communes of France have to bear only a small share of welfare expenditures linked to the economic recession and unemployment. Indeed, it is a national institution that manages the obligatory social security contributions—paid both by employers and employees—and allocates them (pensions, health insurance, and various other allowances). Most additional social welfare expenditures are managed by supracommunal local governments, the departements, which receive an annual subsidy from the municipalities.

TABLE 7.1

Current Receipts of Government as a Percentage of GDP[a]

	1975	1978	1981	1982	1983	1984	1985	1986
Total	37.4	39.5	42.8	43.8	44.6	45.4	45.6	45.1
Central government	16.6	17.3	18.6	18.8	18.3	18.3	18.3	18
Local government	4.3	4.4	4.8	4.9	5.2	5.7	5.9	6
Social security	15.3	16.6	18.3	18.9	19.6	19.8	19.9	19.5

SOURCE: Comptes de la Nation.
a. Obligatory contributions and taxes received by all sorts of governments.

TABLE 7.2

Trends in Local Expenditures and Revenues, 1978-1985, in Real Terms[a]

	1978	1979	1980	1981	1982	1983	1984[b]	1985[b]
Total expenditures	100	105	105	111	118	121	124	129
capital expenditures	100	106	102	110	118	114	114	
current expenditures	100	105	106	112	118	125	131	
Total revenues	100	104	105	109	116	119	125	129
tax revenues	100	103	106	111	119	124	137	
transfers	100	106	109	117	121	125	130	
borrowings	100	95	87	93	105	97	104	
other revenues	100	112	110	107	111	116	121	

SOURCE: Ministère de l'Économie et des Finances.
a. All communes, Paris, *communautes urbaines*, districts, and *syndicats intercommunaux*.
b. Estimate.

- The high proportion of capital expenditures in French commune budgets (about 40% in 1975, 30% in 1986) has given them a lot of room to maneuver. In every country, stabilizing or reducing these expenditures is a very flexible and politically untaxing way to balance the budget at a time when resources are growing scarcer. Stabilizing capital expenditures meant that the growth rate in the communes' expenditures was able to fall from an average of 7% between 1970 and 1975 to approximately 4% from 1975 to 1983.
- For over ten years, pressure for amenities and services has noticeably eased in urban communes, where many realizations were made earlier, especially in the 1950s and 1960s. The number of houses built per year is one of the most important factors in the growth of capital expenditures, and this number fell from 500,000 in 1975 to 330,000 in 1983. The main problem facing local authorities today is not chiefly the quantitative satisfaction of needs for amenities and services, but meeting those needs more satisfactorily, by improving the quality of amenities and adapting them more

TABLE 7.3

Distribution of Local Expenditures and Revenues, 1982-1983 (in percentages)

All expenditures[a]	
Personnel	25.1
Grants, subsidies	14.6
Capital expenses	27.4
Debt	14
Other expenses	18.9
Total	100
current spending[b]	
roads, public lighting	20.3
schools	21.3
sport, culture, leisure	15.5
welfare services and transfers	24.4
others	18.5
Total	100
capital spending[b]	
municipal buildings	11.1
roads, streets	27.6
schools, sport, culture, leisure	34.4
welfare services	4.7
urbanism[c], public transport	17.1
others	5.1
Total	100
Revenues[a]	
tax revenues	34.1
grants, transfers	36.1
borrowings	14.7
other revenues	15.1
Total	100

SOURCE: Ministère de l'Économie et des Finance.

a. 1983, for all communes, Paris, *communautes urbaines*, districts, and *syndicats intercommunaux*.

b. 1982 communes over 10,000 inhabitants only.

c. Expenditures linked to urban planning and urban policy: housing, urban renewal, new quarters planning.

specifically to suit varying requirements. This requires better control over administrative costs and their repercussions on current expenditures.

Strong Effects of Austerity Since 1983

From 1983 onward, the communes' fiscal situation became considerably more difficult. The socialist government began implementing its policy of austerity and announced its intention of reducing income tax, as it was considered an impediment to restructuring the economy. This objective was pursued even more radically by Jacques Chirac's government from 1986 on, but did not entail any measures of constraint as far as the communes were concerned. The central government, faced with the powerful Association of French Mayors, could do no more than make verbal statements recommending moderation in local expenditures if the policy of decentralization was to succeed.

During 1984 and 1985, however, this new policy led to a fall in income for the citizens and to a stagnation in the chief block grant to local authorities (the *Dotation Globale de Fonctionnement*: DGF) that was indexed to VAT revenues (i.e., to domestic consumption). In 1985, overall state transfers increased at a rate lower than inflation for the first time. Municipalities also had to pay the price of the 1983 municipal elections, that is, make up for the low rates of increase in rates that were fixed prior to the elections (1% in real terms, on average) by a substantial increase of 11% in real terms in 1984.

Caught between their election promises and growing discontent on the part of local taxpayers (a poll carried out late in 1983 showed that 63% thought local rates were too high), between a fall in transfers and government orders to play a part and contribute to enforcing *rigueur* (stringency, the left's version of austerity), mayors elected for the first time or reelected saw their room for maneuvering considerably reduced in the climate of the last three years. The right's easy victory in the municipal elections held in March 1983 following the election two years previously of a left-wing president and majority in the parliament changed the political climate. Conservative parties then started a vigorous polemic against government policies, especially decentralization.[2]

Over and above this temporary aggravation regarding fiscal stress and the consequent polemics, there were many symptoms of more profound change in the local political arena. The problem of local finance, and in a newer way the question of local management, became the subject of very wide discussion among local authorities and various experts. Suddenly there were many more publications, talks, and research on these questions (the interest taken in the FAUI survey is revealing). Discussion came about as a result of at least three important facts:

(1) Local authorities now realized that current spending could no longer keep increasing at the same rate as in preceding years (7% in real terms between 1976 and 1983) and that capital expenditures could not continually be reduced or maintained at a level insufficient for renewing or maintaining installations and important infrastructure.

(2) Local authorities were becoming more aware that they would in the future have to rely increasingly on their own strength, that they would be receiving less from central government, and that they alone would be responsible for the political consequences of their decisions. The decentralization process begun in the mid-1970s is probably part of a long-term trend; moreover, the very principles of the reforms carried out by the left are widely approved and do not appear to need any challenge from the conservative government.

(3) The past five years have seen a profound renewal of political ideologies, especially the spread of neoconservative ideas. "Too much government," privatization of government-owned firms and public services, and a return to free-market enterprise have become the slogans and catchphrases of a young generation of right-wing leaders (several of whom were elected mayors in 1983). Several more or less radical attempts at installing a new kind of local management inspired by neoconservatism also played a part in renewing debate on local politics.

Everything points toward the past five years having inaugurated a new phase in the development of local government. According to the FAUI Project survey, a strong majority of mayors (72%) consider they are being faced with quite new problems. These cannot be reduced to problems of fiscal stress, since less than half the mayors state that the fiscal situation of their communes has deteriorated over the past ten years and 45% consider it "satisfactory or fairly satisfactory." But an important key to comprehension of the recent evolution of French local government since the 1970s is the greater institutional and political autonomy gained by the communes.

More Autonomy for the Communes

From the Tutelle to Decentralization

Centralization has long been considered an essential characteristic of French society. Legal, technical, and fiscal tutelage (the *tutelle*) was the state's traditional means of exercising control over the communes. The *préfet* represented and acted for the state in each departement. Other state agents or administrations played an important part in centralization as, for example, the engineers of the Corps des Ponts et Chaussées. Even today, the numerous rural communes are still largely dependent on the expertise and technical capacities of government administrations within the departement.

TABLE 7.4
Chief Administrative Officers' Opinions About 14 Problems (in percentages)

	Very Important	Somewhat Important	Not Very Important	Positive Effect	Negative Effect	Neither Positive nor Negative Effect Don't Know/No Answer
Trend in central state grants for current spending	53	25	23	20	71	9
Limit of increase by central state	52	34	15	4	90	6
Rate of borrowing interest	48	34	19	27	63	24
Local economic situation	43	35	22	23	64	13
Central grants for capital spending	42	26	33	18	66	17
Unemployment	39	30	32	3	82	15
Inflation	38	41	21	8	84	8
Fiscal wealth of the commune	37	32	32	34	50	17
Rising service demands from citizens	31	39	31	14	63	24
Change in demographic situation of the commune	27	25	48	18	47	36
Subsidies from the region	19	21	60	44	27	30
Decentralization reforms	19	30	50	5	60	35
Subsidies from the department	16	27	58	44	30	26
Demands and pressures from municipal employees	5	13	82	7	30	64

SOURCE: FAUI Project survey, CAO questionnaire, 1985-1986.
NOTE: The question was worded as follows: "A lot of constraints make it more difficult to balance local budgets nowadays; could you estimate the importance and the kind of the effects of each following problems?"

But communes were also more or less strictly controlled by Paris, as far as their acts were concerned. In the early twentieth century the Conseil d'Etat, which has supreme administrative jurisdiction, frequently intervened to limit or prevent municipal initiatives. From the 1930s right up until the late 1960s, central government kept on increasing its control over local policies, especially those regarding development and urban planning. Specific subsidies then became an essential way of guiding the communes' investment expenditures (Ashford & Thoenig, 1981).

In portraying the local government, all too often, as being entirely dependent on state administrations, the extent of centralization has undoubtedly been exaggerated. Far from being deprived of political resources, local authorities exercise a role of mutual control with national state civil servants. The *régulation croisée* and the *cumul des mandats* seem to be the mainstays of a kind of *jacobinisme apprivoisé* where central government, apparently all-powerful, has to make allowances for interests of local leaders (Crozier & Thoenig, 1975; Grémion, 1976; Hoffmann, 1974; Worms, 1966). In addition, the 36,000 French communes (33,000 of which have fewer than 2,000 inhabitants) are so heterogeneous that it is necessary to distinguish a local government model that is peculiar to rural communities and small towns — and for this, the analyses by Crozier and Thoenig and by Grémion are particularly relevant — and an urban model, characterized by greater autonomy for the municipalities and greater potential for innovation.

Since the end of the 1960s, the tutelle has become more flexible as the central government has realized that incentive and compromise prove more effective than coercion. This development followed the relative failure of various government attempts to regroup the 36,000 communes within bigger entities as was the case with reforms in other European countries.

From 1974 until 1981, while Giscard d'Estaing was in office, the government abandoned the idea of regrouping the communes and undertook several reforms toward decentralization. But it was from 1981 onward, with Mitterrand's election, that the most important changes took place. From 1981 to 1984 a great number of acts profoundly transformed local administration structures. These reforms, however, have been in relation to the departements and the *regions* more than the communes, especially urban communes: The tutelle by the préfet was abolished even though it had become increasingly supple throughout the 1970s (Meny, 1983), and jurisdictions regarding urbanism and the economy had been widely exercised de facto.

For the communes the fiscal consequences of transferring these responsibilities were very limited (less than 2% of all transferred expenditures to the different local governments). According to the FAUI survey, the implementation of decentralization does not constitute one of the main constraints in the present situation (Table 7.4). Almost half the chief administrative officers consider its consequences slight or negligible; less than 20% consider them very considerable or considerable. It is clear nevertheless that the effects of decentralization on local policies cannot be reduced to these immediate repercussions. These institutional reforms must be put back into the context of the process of change begun before 1981.

A Certain Fiscal Emancipation

Since the 1970s central government has progressively slackened its control over the communes' budget policy. Several successive reforms have modified local taxation, transfers and subsidies, and rules governing access to loans at low interest rates, thereby giving urban municipalities appreciable autonomy.

Nowadays, tax revenues are the main source of local authorities' revenues (approximately 35% of total resources). Most of the communes' tax revenues (85%) come from direct taxation: tax on lands and tax on buildings paid by the owners, *taxe d'habitation* paid by tenants and homeowners, *taxe professionnelle* paid by private business firms and self-employed professional people. Each of these taxes represents a very variable percentage of total direct taxation according to the nature of the economic activities and of the commune inhabitants. On average, the taxe professionnelle accounts for about 50%, the taxe d'habitation for 25%, and the two others taxes on property 25%.

Prior to the reform in 1980, tax rates were calculated by the Finance Ministry according to a very obscure and complicated method. When it came to budget estimations, the municipalities only voted the overall amount of taxation necessary to balance the budget. Since 1980, they themselves decide the annual rate of taxation and can therefore adjust the amounts to be paid each year by all the various categories of taxpayers (property owners, tenants, and businesses). This means that they have considerably wider scope for initiatives, although these are still subject to very strict regulations aimed at avoiding too abrupt changes and at safeguarding the interests of business firms. It is indeed easier, from an electoral point of view, to increase the taxe professionnelle than the other taxes, which affect a much greater number of voters.

Another important change concerns state transfers and subsidies, which account for approximately 30% of communes' total resources. The most important transfer to the communes is the Dotation Globale de Fonctionnement, which covers about 30% of current expenditures. This was initially related to the salaries of private sector, but since 1979 it has consisted of a fraction of VAT revenues automatically shared out among local bodies according to a very complex system designed to reduce progressively the considerable disparities in tax resources that exist from one commune to another. The total amount of the DGF varies more or less according to growth of the GDP, although minimum increases have been guaranteed.

The whole system of specific investment subsidies has also changed radically over the past ten years. Specific subsidies had constituted a basic instrument of central government and its administrations to control the communes' amenities policies. Access to low-interest loans being offered by the Caisse des Dépôts et Consignations (a public body that covers 75% of the communes' loan requirements) was conditional on obtaining a subsidy, and this was awarded only after very lengthy bureaucratic procedures. The fall in these specific subsidies in the early 1970s had provoked discontent among local officials. In 1975 they succeeded in obtaining the creation of a new specific subsidy designed to refund VAT paid by communes on their capital expenditures. The creation of this fund (the FCTVA) largely compensated the fall in specific subsidies.

In 1976 the communes were given the opportunity of negotiating an overall loan from the Caisse des Dépôts et Consignations, thus putting an end to the necessary association between subsidies and loans. Finally, the *Dotation Globale d'Equipement* (DGE) was created in 1983; this was a block grant to be used at the discretion of the communes and designed to replace specific subsidies progressively. The overall amount of the DGE is indexed to the volume of public sector capital spending increases. The proportion received by each commune is calculated on the basis of overall investment expenditures plus other criteria relating to local needs.

Since 1987, municipalities have been allowed more freedom in establishing fees and costs of local public services, the increase of which was previously severely restricted by central government.

Over the past ten years, urban communes have undeniably acquired increasing autonomy regarding institutions and taxation vis-à-vis

central government. Resources transferred to the communes have been
standardized and are virtually all to be used entirely at the discretion of
the communes; augmentation of their overall value is indexed to increased consumption (DGF) or public capital expenditures (FCTVA
and DGE) and is not affected by central government policy decisions
(although the central government can change the rules of the game) and
they are distributed according to complex but fixed criteria that are not
subject to haggling or power struggles.

The price paid for this added autonomy vis-à-vis central government
is a reduction in each local leader's bargaining power and in the
excessive capital of prestige that certain mayors could make out of their
talents for obtaining from central government more subsidies than
others could. Another price to be paid is the servitude of local policies
to fluctuations of the national economy. Finally, it places emphasis on
the municipality-population interface and on local political processes
insofar as local policies are regulated. Denouncing the misdeeds of
"Parisian technocrats" — very convenient scapegoats that they were — is
now a less effective safeguard against possible discontent on the part
of the electorate.

A More Complex and More Active
System of Local Leaders

Local Leaders and Political Parties

French mayors generally have sufficient political and institutional
resources to be considered "strong leaders." Traditionally, mayors were
permitted to hold various local and national offices simultaneously, and
this allowed them to wield personal influence, thus ensuring their
considerable longevity; yet legislation was adopted in 1985 limiting
this cumulation. It was not uncommon for mayors to serve three, four,
or even five consecutive terms of office, each lasting six years.

Over and above the existence of these important local leaders, who
are sometimes prevented from direct running of their town because of
the multiple posts they hold, it seems that local officials have restored
the balance of internal decision-making processes by limiting civil
servants' autonomy. The role of committees[3] consisting of municipal
council members has been extended, and local officials have become
more directly involved in the management of the sectors for which they
are responsible. The FAUI survey shows that as far as the budget is

TABLE 7.5

Budgetary Decision-Making Process (in percentages)

Question: How important was the professional staff as compared to elected officials in
(1) affecting the overall spending level of your budget commune?
(2) allocating funds among departments?

	(1)	(2)
Elected officials play the most important part	72	40
Elected officials and professional staff have about equal part	24	40
Professional staff play the most important part	3	19
(N = 123)		

Question: During the budgetary decision-making process, who makes the most important decisions?

	Mayors (N = 123)	Chief Administrative Officers (N = 155)
Mayor only	11	8
Mayor and a small staff (with the chief administrative officer and one or two deputy mayors	26	33
Mayor and all deputy mayors	51	46
Financial committee	7	7
Municipal council	2	1
Other	2	4
Total	100	100

SOURCE: FAUI Project survey, mayor questionnaire and CAO questionnaire, 1985-1986.

concerned main policy objectives and decisions are determined by elected officials, most often the mayor and his or her deputies. It is likely that local policies are nowadays determined more by deliberate options and political goals than by a simple incrementalist process.

Relations between political parties and the municipalities vary greatly (Dion, 1986; Kukawka, 1983; Lacorne, 1980; Lagroye et al., 1976). In general, local branches of right-wing parties are not very structured or active, although the gaullist party, the RPR, has had its local framework strengthened over the past few years. Socialist representatives often easily become independent regarding local branches of their party, which are mostly incapable of changing the course of municipal policies. The communist party is reputed to exercise strict control over its elected representatives. The party's leaders in fact have supreme control regarding which candidates are designated for the post of

mayor, and the militants have a certain influence over municipal policies. Local situations do, in fact, differ, depending on whether the party
has been strongly established in the commune for a considerable time
or whether the commune has been won over in recent elections and is
run by a pluralist team. In the first instance there is osmosis between
the party and the municipality, and in the second case the links between
the party and local elected officials are much less narrow.

Strange as it may seem, the effect of partisan politics on local
policies, especially budgetary policies, has not been studied very much
in France. Despite the existence of specific political cultures, there is
no evidence that these result in very different local policies. The FAUI
survey affords interesting information on this point.

Municipal Employees

The role of the approximately 750,000 employees in the communes
has also changed. Local bureaucracy was marked by low qualifications,
municipal employees' lack of geographical mobility, lack of training
personnel, and a style of management that was traditional, hierarchical,
and compartmentalized. Over the past ten years, the municipalities
have strengthened their training force by recruiting young staff with
university backgrounds as well as various kinds of specialists: com-
puter experts, economists, management experts, and so on. A few of the
big towns have even recruited members of the *corps préfectoral* (body
of prefects).

Elected municipal officials are naturally obliged to negotiate with
municipal employees' trade unions. Union representation for municipal
employees is fairly strong, as is the case for central government civil
servants, but it varies according to the size of the town. Since its
foundation at the start of the century, the municipal employees' trade
union movement has successfully pursued a strategy of national negotiation with the mayors' association, with the government as arbiter.
The trade unions have progressively obtained statutory guarantees set
out in 1952 legislation that almost bring their status into line with that
of other public employees. Many of the problems affecting municipal
employees are now dealt with during national negotiations.

Mayors have, in fact, considerably less freedom than their U.S.
counterparts, as regards staff management and wages. They do, however, have important leeway with regard to staffing levels, certain
working conditions, and social benefits. Negotiations between elected
officials and municipal employees' trade unions normally lead to agreements, but in the last few years new strategies designed to cut spending

have sometimes led to strikes and conflict when established benefits have been called into question.

According to the FAUI survey, chief administrators, who no doubt feel solidarity with their subordinates, consider that employees' demands have very little influence on budget policy (see Table 7.4). The mayors do not consider municipal employees as a very active pressure group (Table 7.6). In addition, however, mayors see a reduction in staff expenditures as a priority (Table 7.12).

Pressure Groups

In the mid-1960s there was what some referred to as a "boom" in associations. The number of associations being set up each year increased from 12,000 in 1960 to 47,000 in 1983. Estimates now put the number of associations in France at 500,000, and it is calculated that one in three French people belongs to an association of some kind (Passaris & Raffi, 1984).

There is great diversity in these associations and their relations with municipalities are very diverse. Sporting, cultural, and educational associations are broadly financed by the communes, which also provide them access to municipal amenities. They are often run by *notables* (persons of renown) in the case of sports clubs or by people closely related to left-wing parties in the case of educational and cultural associations. They have the strongest influence on municipal policy, as the FAUI survey shows (see Table 7.6).

A second type of association, less influential, has developed in relation to the post-1968 "urban struggles." It extends a movement born in the 1960s (the Groupes d'Action Municipale) that disputed both the old-fashioned way in which the notables governed and the political parties' monopoly of the local political system. This innovative movement was illustrated in Grenoble, where neighborhood committees became important intermediaries between the local inhabitants and the municipality.

Other pressure groups that play a rather important part in local politics, especially in medium-sized towns, are the chambers of commerce and industry. These official bodies represent local employers and often collaborate with the municipalities to finance and manage certain amenities such as airports or industrial parks. They also play an active part in financing public housing, and their influence is often a determining factor in urban planning strategies (Castells & Godard, 1974; Coing, 1977).

TABLE 7.6

Activity and Spending Preferences of 18 Participants and Groups
in City Government Affairs

	Activity Index[a]	Spending Preferences			
		Spend Less	Spend the Same	Spend More	Don't Know/ No Answer
Associations for sport, culture, and leisure	307	1	5	81	13
Municipal council	268	22	28	37	13
Chief administration of departements	264	2	24	60	15
Associations for youth	235	1	11	66	21
The elderly	232	1	37	48	16
Tradesmen's groups	217	27	19	38	16
Users of local services	202	4	9	70	17
Municipal employees' unions	192	1	23	59	18
Neighbors' groups	171	1	11	65	23
Workers' trade unions	159	5	11	49	34
Individual citizens	157	32	20	20	28
Tenants' groups	153	9	11	49	32
Ecologist groups	127	7	12	39	42
Businessmen's institutions	126	18	14	20	49
Central state administrations	123	10	17	24	49
Employers' unions	110	31	11	12	46
Homeowners' groups	105	20	14	19	47
Taxpayers' associations	55	30	4	3	63

SOURCE: FAUI Project survey, mayor questionnaire, 1985-1986.
a. Index: no activity = 0; little activity = 1; some activity = 2; a lot of activity = 3; very important activity = 4.

Table 7.6 shows that, except for the tradesmen's groups, the most active pressure groups are those wanting to increase local expenditures and the less active are those that prefer lower spending. Taxpayers' associations are the least active local groups, according to the mayors. So the local political system seems to be widely characterized by an imbalance in favor of the interest groups that favor more spending.

Citizens

Municipal elections are the main vehicle for expressing citizens' preferences. Participation in local elections, every six years, is high on average (78% in 1983) but varies from about 65% in big towns or suburban communes to 90% in rural communes. In contrast to other countries, the referendum is not a traditional or legal practice in French local politics. Several municipalities, however, have during the last years organized consultation on specific problems or on general municipal policy direction in a populist manner.

Clientelism — mobilization of popular support by elected representatives through a network of personal contacts and rewards — is typical of certain aspects of the way in which municipalities are run. Since the Third Republic it has changed somewhat, but interpersonal relationships continue to play a fundamental role because direct relations between elected officials and citizens enable multiple microadjustments between local policies and individual demands to be made from day to day (Nevers, 1983). The FAUI survey shows that pressures from individual citizens are not considered as very strong by most of the French mayors.

Generally speaking, following the period of the 1950s and 1960s, which was characterized by the municipalities' impoverished political life linked to rapid urbanization, uprooting, and a certain anomie in urban society, the local political system has become more dynamic, especially through more active local leaders and pressure groups. This dynamization was even accelerated by the new constraints of the fiscal stress context and the related policies emerging in the 1980s.

Facing Fiscal Stress: Old and New Strategies

In the present climate of growing fiscal stress, local authorities have a variety of options from which to choose: They can increase resources, cut spending, think up new management methods for making better use of fewer resources, adopt short-term management strategies, "play it by ear," undertake long-term action, or arbitrate among differing social interests.

The FAUI survey of communes with over 20,000 inhabitants carried out in 1985 affords a great deal of information about changes taking place in local politics. As Table 7.7 shows, strategies aimed at increasing resources were those that were most frequently employed, but they are considered less effective, on average, than strategies for reducing spending that have generally been used in more recent times.

TABLE 7.7
Austerity Strategies: General Index About Frequency, Importance,
and Year of Implementation

	Revenue Strategies	Expenditure Strategies	Other Strategies
Frequency[a]	59	53	57
Importance[b]	2.72	2.91	2.80
Year of Implementation[c]	3.61	4.13	3.85

SOURCE: FAUI Project survey, CAO questionnaire, 1985-1986.
a. Index: average percentage of the communes that have applied the strategies.
b. Index: average percentage of responses to the question: how important are the effects of each strategy? Very important = 5; important = 4; rather important = 3; not very important = 2; not important = 1.
c. Index: average percentage of responses to the question about the year of implementation of each strategy. Before 1970 = 1; 1971-1976 = 2; 1977-1979 = 3; 1980-1982 = 4; 1983-1986 = 5. The higher the index, the more recently the strategy was implemented.

Around 1980:
A Fundamental Change in Budget Dynamics

The indicator referring to the year in which the various strategies were used allows a description of how local policies have evolved over the past ten years (see Appendix Table 7.A3). Classification of these strategies that have been regrouped into homogeneous categories gives the following sequence:

(1) increases in local rates and charges

(2) use of additional long-term loans followed by short-term borrowing

(3) intervention in local economic context (urban planning, aid for economic development, repopulation, and so on)

(4) search for additional aid and subsidies

(5) changes in management methods

(6) implementation of other strategies for increasing revenues

(7) implementation of actions to increase productivity

(8) stabilization or reduction of current expenditures

(9) stabilization or reduction of capital spending

(10) stabilization or reduction of staffing levels in the commune

The traditional strategies — increasing local revenues (taxation, charges) or external revenues (subsidies, loans) — predominate until 1980. These strategies are not specifically based on policy formulated to deal with fiscal stress in an economic recession period. They are really

TABLE 7.8

Distribution of the Three Groups of Strategies in Different Periods (in percentages)

	Revenue Strategies	Expenditure Strategies	Other Strategies	
Before 1977	50	20	30	100
From 1977 to 1979	49	20	31	100
From 1980 to 1982	38	30	32	100
1983 and after	29	43	28	100
Total	38	33	30	100

SOURCE: FAUI Project survey, CAO questionnaire, 1985-1986.
NOTE: N = 1,483 (number of strategies for which the year of implementation is available).

characteristic of a period of prosperity similar to that of the 1960s, when increased local fiscal wealth and the growth in household incomes allowed continuous increases of taxation. Local authorities were thus able to finance much of their capital spending through long-term borrowing (over 20-30 years), which is advantageous in an economic climate of growth and moderate inflation.

The limitations of these strategies were seen in the late 1970s. More and more municipalities were obliged to resort to other solutions to balance their budgets. They undertook long-term action to reduce urbanization costs, increase local fiscal wealth, and improve the municipal bureaucracy's productivity. Finally, they initiated a policy of stabilizing or reducing spending, first capital and current spending and then expenditures on staff. During the past five years, expenditure strategies were progressively adopted by the majority of the communes. Until 1980, they accounted for 20% of all strategies implemented, 30% in 1980-1982 and 43% after 1983. Measures to increase resources decreased from 50% to 38%, then to 29%.

These data illustrate quite clearly the profound changes affecting local policies in the early 1980s. A pattern of budget policy based on adjusting revenues to expenditures had persisted until the late 1970s. Increasing expenditures from whatever sources were the catalyst to local budget dynamics. The level of resources needed was conditioned by the level of expenditures. Now the process whereby increasing taxation is just a consequence of responses to changing needs and demands for expenditures has been gradually replaced by a new logic based on the scarcity of available revenues. In most communes local authorities predetermine a tax level relative to other expected revenues and this becomes the major political consideration when the budget is being planned. Increases in spending have to be adapted to this level. As the financial director of a town vividly put it: "We used to do additions and now we do subtractions."

What Is Now (Almost) Impossible: Augmenting Revenues

One of the oldest and most frequent strategies was to increase local rates, which accounted for approximately 35% of the communes' total resources. It was deemed the most effective policy tool and, in fact, made it possible for communes to meet constantly increasing local expenditure costs from the early 1960s on.

For the last few years, however, reducing or stabilizing the local rates has been the main objective of mayors, who are very sensitive to growing discontent among their electorate. This was one of the strongest findings of the FAUI survey. Only 15% of mayors think they can still increase the tax burden, 38% state that it must be stabilized, and 44% say that it must be reduced (see also Table 7.12). It should nevertheless be pointed out that there is a contrast between the objectives of the great majority of the mayors and what they do in practice: Whereas many mayors (especially on the right) were elected or reelected in 1983 on the strength of their promises to reduce taxation, very few of them have been able to do it.

The disparity between stated objectives and actual practice shows the increasing problems the mayors must solve, knowing as they do that other revenue strategies do not provide effective alternatives to local taxation. For example, another strategy widely used in the past was the search for additional aid from central government and from other local bodies (the region and the departement). But now this is fairly ineffective because indexing and standardizing block grants from the state leaves local officials little room to maneuver and the departements and regions are themselves facing increasingly urgent fiscal problems.

Previously, municipalities relied largely on borrowing to finance their capital spending, but the level of debt incurred following a "boom" in local investment during the 1960s, spiraling interest rates, and, more recently, the reduction in the rate of inflation all induced municipalities to cut back on their borrowing. In the chief administrative officers' opinion, the excessively high cost of loans is one of the most important fiscal constraints on balancing local budgets (see Table 7.4); furthermore, a large majority of mayors (71%) is opposed to increasing borrowing (Table 7.12).

TABLE 7.9

Revenue Strategies: Use and Relative Importance (ranked according to importance)

	Using[a]	Index[b]
Increase local taxes	78	56
Increase long-term borrowing	60	44
Draw down surplus	64	41
Increase user fees and charges	77	38
Request additional subsidies from region	65	35
Seek new local revenue sources	64	35
Request additional subsidies from departement	63	33
Request additional or new state grants	60	33
Improve the use of paying services	56	33
Sell some assets	43	29
Increase short-term borrowing	41	31
Suppress exemption of fees on certain services	36	20

SOURCE: FAUI Project survey, CAO questionnaire, 1985-1986.
a. Percentage of the communes that have applied the strategy.
b. This index is a weighted average showing the importance of the efficiency of each strategy according to the chief administrators. The formula of the index is as follows:

$$\frac{(5 \times A) + (A \times B) + (3 \times C) + (2 \times D) + (1 \times E)}{5 \times (A + B + C + D + E + Y)} \times 100$$

where A = number of "very important" responses; B = important; C = rather important; D = not very important; E = not important; Y = not used and nonresponses.

More than two-thirds of the communes have increased user fees for local services, but this has had very little effect because of the local public services' freeze applied by the central government in the past few years. Chief administrators see the freeze as the biggest fiscal constraint on communes (Table 7.4) and nearly three out of every four mayors are in favor of higher increases in user fees (Table 7.12). These facts show that in the present economic climate, where the burden of taxation has reached the point where it is becoming increasingly intolerable, mayors must urgently reconsider their strategies regarding user fees for services. The current policy (applied since December 1986) of liberalizing prices through a progressive suppression of the freeze will give them more autonomy to decide whether to keep user fees low or set them as close as possible to the actual costs of providing services. Arbitrating between users and taxpayers is not easy from the political point of view, however; in any case, even a substantial increase in fees cannot compensate for reduced or stabilized tax revenues. User fees account for less than 12% of all revenues, and a substantial proportion of local public goods are nonseparable goods.

Two other strategies can help make local services more "profitable": 36% of surveyed communes increased the use of services with fees, and 43% decided to charge for services that were previously free. Both strategies (especially used after 1980) are strong indications that a new policy is emerging, although the results are still difficult to ascertain. Charging for services that were previously free symbolizes a new policy stance in favor of taxpayers; action taken to encourage people to make more use of services is particularly designed to make them more "visible" to the electorate.

What Must (Unfortunately) Be Done: Reduce Spending

For some years now a reduction or slowing down of local spending increases has become a necessary objective, as it is now impossible to procure additional revenues. More than 80% of mayors are in favor of reducing current expenditures and also staff costs, even though they are aware of the difficulties in implementing such a policy (see Table 7.12). The findings of the FAUI survey clearly reveal three different steps of the retrenchment policy French communes are beginning to implement.

In a first step the great majority of communes, more than 75%, have (since 1980 in particular) applied strategies of reducing capital expenditures by extending the time allotted for new projects or withdrawing them altogether and reducing general costs, including supplies, and 87% of communes have slowed down or halted new municipal job creation.

The second step was taken more recently (after 1983) by about half of the communes. They applied more radical strategies, such as across-the-board reductions in current expenses or selective cuts in the budget allocated for certain services, cuts in expenditures for depreciation, and renewing or maintaining stocks. They have also begun reducing staffing levels in their communes by not replacing employees who retire.

Third, very few communes have decided to lay off personnel or close down certain services; whenever this has been done, it has had very limited effects. Less than a third of the municipalities have reduced their local subsidies, mainly to the many cultural, sports, educational, or social organizations that make up the most active pressure groups, as we have already seen.

It is difficult to measure the exact extent of the effects of these spending-reduction strategies. By far, the most important of these were

TABLE 7.10
Expenditure Strategies: Use and Relative Importance
(ranked according to importance)

	Using[a]	*Index*[b]
Halt creation of additional jobs, freeze hiring	87	56
Reduce capital expenditures	77	53
Seek better conditions for purchases	90	52
Reduce operating expenses	86	51
Defer or eliminate programs	78	47
Institute cuts in all departments	52	39
Reduce work force through attrition	56	37
Defer or reduce maintenance of capital stock	46	36
Institute cuts in only some departments	46	30
Decrease turnover of working stock	45	28
Defer some payments to next year	30	23
Institute cuts in subsidies allocated by the commune	31	22
Close some services	16	14
Lay off administrative or service personnel	6	12

SOURCE: FAUI Project survey, CAO questionnaire, 1985-1986.
a. Percentage of the communes that have applied the strategy.
b. This index is a weighted average showing the importance of the efficiency of each strategy according to the chief administrators. See note b, Table 7.9, for index formula.

cuts in capital spending, which represents 27% of commune expenditures and those concerning municipal personnel (about 25% of total expenditures). In 1984 total expenditures in real terms had fallen in 34% of the towns surveyed; capital spending was lower in 54% of these towns, but current expenses were lower in only 8% and staff costs lower in 12%. In 1985, current expenditures had increased at a rate lower than that of inflation in 14% of the communes surveyed; 25% of the communes also reported that personnel expenditures increased at a rate lower than the inflation rate.

Innovating: Consensus and Conflict Strategies

The political difficulties caused by implementing most of these revenue and spending strategies have prompted most municipalities to consider long-term strategies to improve productivity and to stimulate economic development. Almost all the communes established energy-saving programs (subsidized by the central government) and introduced new technology, especially computers. Data processing was adopted progressively, according to the size of the communes; mayors

TABLE 7.11

Other Strategies: Use and Relative Importance (ranked according to importance)

	$Using^a$	$Index^b$
Set up energy-saving program	91	58
Improve productivity by adopting new technologies	94	57
Improve personnel training	73	44
Increase control of employees' work performance	61	35
Institute new methods of management, expert evaluation	39	24
Support new activities and jobs	94	53
Increase urban planning	54	37
Stop drop of population; stimulate growth	58	33
Support voluntary activities	43	24
Shift responsibilities to other institutions	34	21
Contract out with private sector	31	21
Institute municipalization (take back a contracted service)	10	8

SOURCE: FAUI Project survey, CAO questionnaire, 1985-1986.
a. Percentage of the communes that have applied the strategy.
b. This index is a weighted average showing the importance of the efficiency of each strategy according to the chief administrators. See note b, Table 7.9, for index formula.

and chief administrators see it as the foremost innovation of the past ten years. Three out of every four communes began improving professional training of their employees and 61% stepped up their inspections of the work of their personnel. It is obviously difficult to measure real productivity gains resulting from these various acts; in the chief administrators' opinions, only the energy-saving programs and the use of computers have had significant effects.

All municipalities have begun taking action to encourage new economic activities in their territory. There are two reasons for this: to reduce unemployment and to increase the town's tax revenues. These actions are varied and range from installing infrastructure and costly amenities to advertising campaigns to "sell" the town. Far fewer municipalities have adopted any long-term strategy for reducing urbanization costs and reversing population decline.

These productivity and economic development strategies are likely to meet with wide social approval. This is not the case for other approaches involving alternative management methods for local services: privatization or contracting out, "debudgetizing" (isolating one or several services out of the general municipal budget into one or several separate budgets), "municipalization" (the opposite of privati-

TABLE 7.12

Preferences of Mayors About 13 Austerity Strategies (in percentages)

	All Mayors			Left Mayors			Right Mayors		
	(1)	*(2)*	*(3)*	*(1)*	*(2)*	*(3)*	*(1)*	*(2)*	*(3)*
Reduce investment programs	60	34	7	56	31	13	63	33	4
Reduce current expenditure	79	14	7	75	17	8	85	10	6
Maintain or reduce work force	83	10	7	79	10	10	88	7	6
Freeze wages and salaries	13	78	9	8	81	10	17	75	8
Increase user fees	73	21	7	63	27	10	81	15	4
Increase local taxes	11	83	7	19	71	10	6	90	3
Pressure central government for further grants	85	8	7	81	10	8	86	7	7
Ask region and departement to fund projects jointly	85	8	7	75	13	13	90	6	4
Increase borrowing	22	71	7	15	71	8	26	68	6
Eliminate or cut certain services	33	60	7	21	71	8	43	50	7
Contract out to private sector	39	51	10	10	81	8	60	29	11
Institute municipalization (take back contracted services)	16	63	22	27	48	25	7	72	21
Support voluntary activities	69	21	11	69	17	15	69	22	8

SOURCE: FAUI Project survey, mayor questionnaire, 1985-1986.
NOTE: Question: What is your opinion about the following strategies for dealing with the financial problems of the communes? Responses: (1) very much in favor and in favor; (2) not in favor and very much not in favor; (3) don't know/no answer.

zation), and, to a lesser extent, appeals for volunteers. Over half the communes have not implemented a single one of those strategies.

Given the ideological and political climate in France, where nationalization policies implemented in 1981 by the socialist government have been followed by reprivatization policies since 1986, one can imagine that any decision to municipalize or privatize a local service must be of some political importance, even if, as often happens, it is only a question of pragmatism. Indeed, these are the strategies that most clearly differentiate left-wing municipalities from those of the right. According to the FAUI survey, only 15% of communist and socialist municipalities have contracted out one or more services, in comparison with 45% of right-wing ones. This contrast is even more obvious when mayors are questioned about their strategy choices: No communist mayor and only 13% of socialist mayors were in favor of privatization, against 70% of RPR mayors (Jacques Chirac's party) and 51% of UDF mayors (the nongaullist right).

Conclusions

Commune administration is one of the most stable institutions in French society. This does not mean it is conservative; on the contrary, it is partly because it has progressively adapted that it has managed to guarantee its stability. Another reason for its stability is the ability of local officials (who are well skilled at manipulating an extremely centralized government) to direct the discontent of voters toward central government.

Recent reforms carried out by the left-wing government can be traced back to demands made in the 1960s by new local elites from the booming middle classes. These new local elites were the most active sections of the socialist party's electoral base and in the municipal elections in 1977 they were victorious in almost one-third of urban communes, especially in medium-sized towns (having fewer than 100,000 inhabitants); they encouraged an autonomy process in these communities, mainly vis-à-vis central government, a process long since under way in large towns. Support obtained from these new local leaders contributed to Mitterrand's election as president in 1981 as well as the government's pursuit of new policies of nationalization and decentralization over the following two years.

But the effects of the economic recession, the government austerity policy implemented since 1983, and the national victory of the right-wing parties in 1986 changed the meaning of decentralization. Henceforth, it was a question of government's "decentralization of austerity" — that is, make local governments manage the social consequences of the crisis, encourage them to take action to develop their local economies, and, especially, make them contribute to the national policy of reducing compulsory taxation. The last, of course, presupposes implementing measures to cut local spending.

One can understand that circumstances have become much more difficult for local authorities, who have, until recently, been relatively unaffected by fiscal stress. In some ways, local governments can make decisions more freely, but these decisions are less popular and often are even totally unpopular. As we have seen, however, municipalities have recently embarked on a new path of stabilizing or reducing current expenditures and staff expenditures. The spreading implementation of these strategies marks a basic turning point in local policies.

At this point of the analysis of the FAUI data, it is difficult to determine which factors are behind the choice of one or another strategy or group of strategies. We shall comment only that political affiliation seems to play an important part; it clearly differentiates left-wing municipalities attached to the public management of local services and some of the right-wing municipalities that want to break with what they call "municipal socialism." Even if party differences are not so important when it comes to applying specific spending-reduction strategies, their social and political significance is best understood by situating them in the context of these two contradictory goals: to preserve as best they can the basics of the "welfare commune" or to promote another model of local management, as some newly elected mayors are currently attempting to do.

Appendix: Austerity Strategies

TABLE 7.A1

Frequency of Usage (ranked according to frequency)

	Using[a]	Index[b]
(C12) Improve productivity by adopting new technologies	94	57
(C21) Support new local activities and jobs	94	53
(C11) Set up energy-saving program	91	58
(B03) Seek better conditions for purchases	90	52
(B01) Freeze hiring	87	56
(B04) Reduce operating expenses	86	51
(A01) Increase local taxes	78	56
(B05) Defer or eliminate programs	78	47
(B02) Reduce capital expenditures	77	53
(A04) Increase user fees and charges	77	38
(C13) Improve personnel training	73	44
(A05) Request additional subsidies from region	65	35
(A06) Seek new local revenue sources	64	35
(A03) Draw down surplus	64	41
(A07) Request additional subsidies from department	63	33
(C14) Increase control of employees' work performance	61	35
(A02) Increase long-term borrowing	60	44
(A08) Request additional or new state grants	60	33
(C22) Stop drop of population	58	33
(A09) Improve use of paying services	56	33
(B07) Reduce work force through attrition	56	37
(C22) Increase urban planning	54	37
(B06) Institute cuts in all departments	52	39
(B08) Defer or reduce maintenance of capital stock	46	36
(B09) Institute cuts in only some departments	46	30
(B10) Decrease turnover of working stock	45	28
(C31) Support voluntary activities	43	24
(A11) Sell some assets	43	29
(A10) Increase short-term borrowing	41	31
(C15) Institute new methods of management, expert evaluation	39	24
(A12) Suppress exemption of fees on certain services	36	20
(C32) Shift responsibilities to other institutions	34	21
(B12) Institute cuts in subsidies allocated by the commune	31	22
(C33) Contract out with private sector	31	21
(B11) Defer some payments to next year	30	23
(B13) Close some services	16	14
(C34) Institute municipalization (take back contracted services)	10	8
(B14) Lay off administrative or service personnel	6	12

SOURCE: FAUI Project survey, CAO questionnaire, 1985-1986.
NOTE: A = revenue strategies; B = expenditure strategies; C = other strategies.
a. Percentage of the communes that have applied the strategy.
b. Index: very important = 5; important = 4; rather important = 3; not very important = 2; not important = 1; not used = 0.

TABLE 7.A2

Relative Importance (ranked according to importance)

	Using[a]	Index[b]
(C11) Set up energy-saving program	91	58
(C12) Improve productivity by adopting new technologies	94	57
(B01) Freeze hiring	87	56
(A01) Increase local taxes	78	56
(C21) Support new local activities and jobs	94	53
(B02) Reduce capital expenditures	77	53
(B03) Seek better conditions for purchases	90	52
(B04) Reduce operating expenses	86	51
(B05) Defer or eliminate programs	78	47
(C13) Improve personnel training	73	44
(A02) Increase long-term borrowing	60	44
(A03) Draw down surplus	64	41
(B06) Institute cuts in all departments	52	39
(A04) Increase user fees and charges	77	38
(C22) Increase urban planning	54	37
(B07) Reduce work force through attrition	56	37
(B08) Defer or reduce maintenance of capital stock	46	36
(A05) Request additional subsidies from region	65	35
(A06) Seek new local revenue sources	64	35
(C14) Increase control of employees' work performance	61	35
(A07) Request additional subsidies from departement	63	33
(A08) Request additional or new state grants	60	33
(C22) Stop drop of population	58	33
(A09) Improve use of paying services	56	33
(A10) Increase short-term borrowing	41	31
(B09) Institute cuts in only some departments	46	30
(A11) Sell some assets	43	29
(B10) Decrease turnover of working stock	45	28
(C31) Support voluntary activities	43	24
(C15) Institute new methods of management, expert evaluation	39	24
(B11) Defer some payments to next year	30	23
(B12) Institute cuts in subsidies allocated by the commune	31	22
(C32) Shift responsibilities to other institutions	34	21
(C33) Contract out with private sector	31	21
(A12) Suppress exemption of fees on certain services	36	20
(B13) Close some services	16	14
(B14) Lay off administrative or service personnel	6	12
(C34) Institute municipalization (take back contracted services)	10	8

SOURCE: FAUI Project survey, CAO questionnaire, 1985-1986.
NOTE: A = revenue strategies; B = expenditure strategies; C = other strategies.
a. Percentage of the communes that have applied the strategy.
b. Index: very important = 5; important = 4; rather important = 3; not very important = 2; not important = 1; not used = 0.

TABLE 7.A3

Year of Implementation (ranked according to year of implementation)

		Index[a]	1983-1986	1980-1982	Before 1980
			Year of Implementation		
(B01)	Freeze hiring	473	79	16	5
(B07)	Reduce work force through attrition	460	80	11	8
(B06)	Institute cuts in all departments	442	67	17	17
(B09)	Institute cuts in only some departments	438	62	34	10
(C11)	Setting up energy-saving program	435	52	36	12
(B12)	Institute cuts in subsidies allocated by the commune	432	77	5	19
(B04)	Reduce operating expenses	431	58	25	16
(B02)	Reduce capital expenditures	430	60	19	21
(C14)	Increase control of employees' work performance	429	63	17	20
(C15)	Institute new methods of management, expert evaluation	424	62	24	14
(A12)	Suppress exemption of fees on certain services	419	55	23	22
(B05)	Defer or eliminate programs	411	60	19	21
(C31)	Support voluntary activities	410	55	25	20
(B13)	Close some services	400	58	17	25
(A03)	Draw down surplus	395	44	29	28
(A06)	Seek new local revenue sources	385	37	37	26
(C13)	Improve personnel training	384	49	24	27
(B10)	Decrease turnover of working stock	382	36	41	23
(A09)	Improve use of paying services	382	53	12	36
(B08)	Defer or reduce maintenance of capital stock	381	33	41	22
(C32)	Shift responsibilities to other institutions	372	45	21	45
(C34)	Institute municipalization (take back contracted services)	370	30	30	40
(C33)	Contract out with private sector	370	46	17	38
(B14)	Lay off administrative or service personnel	370	44	22	33
(C21)	Support new local activities and jobs	369	38	24	39
(B11)	Defer some payments to next year	369	30	35	34
(C22)	Increase urban planning	365	42	15	42
(A10)	Increase short-term borrowing	364	32	21	47
(C12)	Improve productivity by adopting new technologies	362	25	37	38
(B03)	Seek better conditions for purchases	359	46	17	36
(A11)	Sell some assets	356	49	8	43
(A07)	Request additional subsidies from departement	356	37	27	37
(A08)	Request additional or new state grants	355	27	29	44
(A05)	Request additional subsidies from region	352	32	30	38

TABLE 7.A3

(continued)

| | | Year of Implementation | | |
| | | | | |
		Index[a]	1983-1986	1980-1982	Before 1980
(A02)	Increase long-term borrowing	339	23	25	53
(C22)	Stop drop of population	333	31	18	51
(A01)	Increase local taxes	330	30	18	53
(A04)	Increase user fees and charges	298	25	17	57

SOURCE: FAUI Project survey, CAO questionnaire, 1985-1986.
NOTE: A = revenue strategies; B = expenditure strategies; C = other strategies.
a. Index: before 1971 = 1; 1971-1976 = 2; 1977-1979 = 3; 1980-1982 = 4; 1983-1986 = 5.

Notes

The FAUI Project research has been conducted in France since 1984 by Jeanne Becquart-Leclercq (University of Lille), Vincent Hoffmann-Martinot (CNRS, Centre d'Etude et de Recherche sur la Vie Locale of Bordeaux), and Jean-Yves Nevers (CNRS, Equipe de Recherche Modes de Production et Société de Toulouse), with support of the Research Department and the Housing Department. Mayors and chief administrators of all cities over 20,000 were surveyed in 1985. Questions were very similar to those of the American questionnaires. Furthermore, several case studies were conducted in 1986. Results presented in this text involve data from the first general analyses.

1. The main units of French local government (without the abroad territories) are *communes* (36,512), *departements* (96), and *regions* (22).

2. From 1982 and the adoption of the March 2 law to 1986, the left government implemented a large decentralization reform: suppression of the a priori tutelle, prefects replaced by local officials at the head of the departements and regions, and transfer of many functions from central government to local authorities.

3. Normally, in each municipality, committees are constituted by council members for different municipal areas: finances, urbanism, culture, and so on.

References

Ashford, D., & Thoenig, J.-C. (Eds.). (1981). *Les aides financières de l'etat aux collectivités locales en France et à l'étranger*. Paris: Litec.

Biarez, S., & Souchon-Zahn, M.-F. (1987). Des associations actives dans les grandes villes. *Project, 203*, 57-72.

Caisse des Dépôts et Consignations. (1986). *Tableau de bord des finances locales: Statistiques commentees (1970-1984)*. Paris: Editions du Moniteur.

Castells, M., & Godard, F. (1974). *Monopolville*. Paris: Mouton.

Coing, H. (1977). *Des patronats locaux et le défi urbain*. Paris: Editions du CRU.

Crozier, M. (1965). *Le phénomène bureaucratique*. Paris: Le Seuil.

Crozier, M., & Friedberg, E. (1977). *L'acteur et le système*. Paris: Le Seuil.

Crozier, M., & Thoenig, J.-C. (1975). La régulation des systèmes organisés complexes: Le cas du système de décision politico-administratif local en France. *Revue Française de Sociologie, 16*(1), 3-32.

Dion, S. (1986). *La politisation des mairies.* Paris: Economica.

Grémion, P. (1976). *Le pouvoir périphérique.* Paris: Editions du Seuil.

Grémion, P. (1978, June). Les associations et le pouvoir local. *Esprit, 18,* 19-31.

Hoffmann-Martinot, V., & Nevers, J.-Y. (1985). Les maires urbains face à la crise. *Les Annales de la Recherche Urbaine, 28,* 121-132.

Hoffmann, S. (1974). *Essais sur la France: Déclin ou renouveau.* Paris: Le Seuil.

Jobert, B., & Sellier, M. (1977). Les grandes villes: Autonomie locale et innovation politique. *Revue Française de Science Politique, 27*(2), 205-227.

Kesselman, M. (1967). *The ambiguous consensus: A study of local government in France.* New York: Alfred A. Knopf.

Kesselman, M. (1974). Political parties and local government in France: Differentiation and opposition. In T. N. Clark (Ed.), *Comparative community politics* (pp. 111-138). New York: John Wiley.

Kobielski, J. (1975, June 20-24). *Tendance politique des municipalités et comportements financiers locaux.* Paper prepared for presentation at the Conference on Politics, Policy and the Quality of Urban Life, Bellagio.

Kukawka, P. (1983). Les partis politiques français et la structuration du pouvoir local. In A. Mabileau (Ed.), *Les pouvoirs locaux à l'épreuve de la décentralisation.* Paris: Pédone.

Lacorne, D. (1980). *Les notables rouges.* Paris: Presses de la FNSP.

Lagroye, J., et al. (1976). *Les militants politiques dans trois partis français: Parti Communiste, Parti Socialiste, Union des Démocrates pour la République.* Paris: Pédone.

Mabileau, A., et al. (1983). *Le personnel communal, les élus et l'administration d'état dans la région Aquitaine (ressources humaines et aménagement de l'espace).* Bordeaux: C.E.R.V.L.

Mény, Y. (1983). Pouvoir administratif d'état et collectivités territoriales. In A. Mabileau (Ed.), *Les pouvoirs locaux à l'épreuve de la décentralisation* (pp. 113-126). Paris: Pédone.

Nevers, J.-Y. (1983). Du clientélisme à la technocratie: Cent ans de démocratie communale dans une grande ville, Toulouse. *Revue Française de Science Politique, 33*(3), 428-454.

Passaris, S., & Raffi, G. (1984). *Les associations.* Paris: La Découverte.

Sorbets, C. (1983). Est-il légitime de parler d'un présidentialisme municipal? *Pouvoirs, 24,* 105-115.

Tarrow, S. (1977). *Between center and periphery: Grassroots politicians in Italy and France.* New Haven, CT: Yale University Press.

Wolman, H., & Goldsmith, M. (1985, July 15-20). *Local government fiscal behaviour and intergovernmental finance in a period of slow national growth: A comparative analysis.* Paper presented at the International Political Science Association World Congress, Paris.

Worms, J.-P. (1966). Le préfet et ses notables. *Sociologie du Travail, 3,* 249-276.

8

Analyzing Determinants of Fiscal Policies in French Cities

RICHARD BALME

The Analytical Framework: Environment, Leaders, and Fiscal Strategies

The preceding chapter set out the French context of local government and presented an initial evaluation of the use of these strategies. The purpose of this chapter is to test a general model explaining the determinants and constraints shaping fiscal policies in French cities. The core model was first proposed by Clark, Burg, and de Landa (1984); several authors in this volume have refined the model for use in non-American settings.

The dependent variables are the 38 fiscal strategies selected by city governments, classified into three major groups: strategies to reduce expenditures, strategies to increase revenues, and strategies to improve productivity. For the last two, we especially focused on strategies to increase taxes and strategies of privatization (dependent variables are presented in Appendix A of this chapter). These strategies are reported by city officials to have been used over the past ten years by their city governments. We distinguish these choices from actual policy outputs and outcomes (Levy, Melstner, & Wildavsky, 1974), since their effective implementation remains to be tested with analyses of budgetary data. Rather, we consider their selection to indicate an authoritative decision by local officials.

This study introduces a new local policy typology in place of the familiar dimensions such as allocative, redistributive, or developmental effects of decisions (Lowi, 1969; Peterson, 1981). This new typology emphasizes policy effects on the importance and the efficien-

213

cy of government activities within the city: Strategies of retrenchment account for the limitation of government size; strategies to increase resources influence the maintenance or development of the public sector; and orientations toward productivity contribute to improved efficiency. Although these types of strategies are not exclusive,[1] we focus here on specifying the different factors leading to and constraining different policy choices.

The main independent variables cover three sets of factors: socioeconomic situations of cities, local political processes, and leaders' characteristics (independent variables are presented in Appendix B of this chapter). We first consider general contextual variables provided by census data for 1982 and, more specifically, population size, unemployment, and foreign population[2] and their respective changes from 1975 to 1982. The relationship of migration and ethnic composition with local fiscal policies is frequently reported in American analyses (Clark & Ferguson, 1983; Shefter, 1985); here, we consider their importance for French cities' politics and policies. We also use indicators of social stratification, in terms of age and occupational positions, to construct several indices for upper-, middle-, and low-income classes. These reflect the effects of population characteristics on policy decisions.

To consider economic dimensions, we build four ratios of fiscal strain by relating different budget indicators to the local "fiscal potential," a measurement of city wealth.[3] The relative economic welfare of the city is thus related to taxes per capita, debt per capita, and total spending for the same year (1982). The form and intensity of fiscal strain are expected to have differential influence on strategies for coping with austerity.[4]

These environmental characteristics taken into account, the focus turns to political processes. The first ratio is a measure of relations between local and central government in the budgetary process, relating the level of national grants to local governments to the total amount of local spending. Such relations are expected to shape local policies: Dependence on national grants, in association with decentralization reforms, can be a strong incentive to mayors for retrenchment.

We also expect local bureaucracy to be a major constraint in the selection of innovative decisions; a rough measure of its importance is provided by the number of municipal employees per capita. Other political variables involve mayoral leadership. The political affiliation of the mayor is used, after recoding, as an ordinal variable representing a continuum from the left to the right. However, this affiliation may vary from a simple partisan label to a deep involvement in partisan organizations, in the case of France strengthened by the *cumul des mandats* (cumulation of elected positions at different level of government). Therefore, we built an index of the partisan integration of mayors, summing their scores for their responsibilities within parties, the role of the party during their election campaign, and the frequency of their disagreements with party members.

These variables allow us to evaluate the influence of partisan affiliation on fiscal strategies, in terms of both their nature and their intensity. They do not, however, refer to the actual personal influence of the mayor on fiscal policy decisions. To provide a measure of this dimension, we use two different indices. The first refers to the decision process, measuring the role of elected officials vis-à-vis appointed municipal managers in determining both the total amount of spending and its distribution among services. This index provides a measure of incrementalist decision processes (Wildavsky, 1964), as opposed to more active involvement of elected officials in the budgetary process.[5] The specific influence of mayors on policy outcomes is measured by a second index reflecting their preferences concerning the different types of fiscal strategies described above (favor/oppose). We also constructed indices of fiscal liberalism and social conservatism, using mayors' evaluations of specific social issues.

If leaders matter, we need to understand when and how leadership makes a difference. We measure both citizen responsiveness and group responsiveness to characterize relations between leaders and their constituents. Mayors were asked, "How often do municipal policies respond to preferences of individual citizens?"; this same question also was asked regarding 18 categories of groups, including municipal employees' organizations, business groups, labor unions, and voluntary associations. We developed several group indices, including and excluding groups within the city hall (employees, managers, and the council); we also distinguished groups with expected high-spending preferences (such as municipal employees and leisure groups) from those with expected low-spending preferences (e.g., business groups and homeowners). However, these different indexes are highly correlated and produce very similar results; thus, unless specified, we use the sum of scores for all categories of groups. Similarly, we test the effects of groups' and citizens' activity, as perceived by mayors, for the same categories. Finally, we analyze the effects of dynamic leadership through the discrepancy between the mayors' decisions and their perceptions of their constituents' expectations ("How often do you take positions against your constituents?").

Finally, following Clark and Ferguson (1983), the basic dimensions of political culture analyzed here include fiscal and social preferences of leaders and their responsiveness to groups and citizens. The rightist political culture combines conservatism on fiscal and social issues with responsiveness to citizens rather than to organized groups. The leftist political culture is symmetrical with social and fiscal liberalism and responsiveness to groups more than to citizens. New populists combine fiscal conservatism with social liberalism, and stress responsiveness to citizens compared to groups.

Origins of Retrenchment: Strategies to Reduce Expenditures

The first regression analysis shows that strategies to reduce expenditures are primarily determined by four main factors: fiscal strain stemming from debt, population size, foreign population and its growth, and leaders preferences' concerning this specific type of strategy (see Appendix C, Equation 1). Some of these relations are rather obvious: Leaders' preferences for retrenchment strategies are directly related to their selection of these strategies, although the relationship between retrenchment and fiscal strain resulting from debt, rather than taxes or spending, implies a more managerial than political origin for such strategies.

Other relationships suggest less immediate explanations: Population size could reflect the importance of fiscal problems but appears here actually to limit fiscal strain (see Appendix C, Equation 2). Therefore, population size does not directly influence fiscal policies through the economic conditions of city government, linked for instance to economies or diseconomies of scale. Rather, population size affects the political feasibility of fiscal retrenchment; as cities grow in size, the level of spending per capita and number of municipal employees per capita increases (respectively, $r = +.247$ and $r = +.194$). Similarly, the larger the city, the larger the upper class ($r = .198$ with managers and professionals), the older the population ($r = .172$), and the more the city government remained to the right in the last election ($r = .130$). If population size does not generate fiscal strain, it does affect the scope of strategies used to reduce expenditure, as well as the social conditions of support for such policies.

The political salience of social conditions is also signified by the importance of foreign population for retrenchment strategies.[6] Two points must be emphasized here. First, the foreign-born population does not participate in elections and, therefore, affects policies only indirectly. These social groups thus cannot be seen as pressure groups within local political processes, especially since these ethnically heterogeneous interests remain at a very low level of organization in French cities. Second, the foreign-born population is basically concentrated in low-income and low-occupational categories and is also more exposed to unemployment than French citizens. Given these conditions, preferences of foreign-born groups are difficult to evaluate; we can, at least, identify their interests in maintaining or developing governmental support. And here we see a contradictory relationship of foreign population vis-à-vis its own interests — the larger the growth of foreign population in the city, the more expenditure-reducing strategies have been selected by leaders. Two alternative explanations are likely: Since the foreign population does not affect policy outputs through political activity, these effects either stem from the reactions of other social groups to their importance in the city or reflect the importance of other economic features. We consider these two arguments in this section, starting with economic and social determinants of retrenchment.

Structural Determinants of Retrenchment

What are the social conditions of retrenchment? If social groups support different policies, or if they exhibit different preference patterns, the relative importance of these groups within the city is expected to be reflected in local strategies of retrenchment. Our initial assumption was that large upper and upper-middle classes were more likely to favor a reduction of local government expenditures, and low-income classes to oppose it. Tests of this relationship, employing distinct and cumulative categories for each of these classes, contradict this view. None of the measures of middle class display any significant relationship with selection of retrenchment strategies, nor are upper-middle-class groups (managers and professionals) associated with these decisions. Broader categories such as employees or workers, considered separately or cumulatively, also show no significant relationship with retrenchment strategies. Two other relationships, however, are significant and distinctive. The greater the presence of traditional upper-class groups (businessmen, merchants, and shopkeepers), as well as the larger the proportion of elderly in the population, the less likely these retrenchment strategies would be selected by local governments. This negative relation is surprising, given the importance of these groups in the rightist electorate and the fiscal conservatism generally attributed to their preferences. Furthermore, there is a positive relation between a larger proportion of popular classes (workers, employees, and low middle class, cumulated) and the selection of retrenchment strategies. In short, the larger the foreign population and the larger the popular classes, the more likely expenditures have been reduced. In contrast, the larger the traditional upper class and the older the population, the less likely these strategies have been selected by leaders.

As these independent variables are intercorrelated, it is difficult to separate their effects and to isolate the nature of the policy determinant (population age rather than occupational or ethnic characteristics, for instance). Because of the ideological salience of the issue, we might consider foreign stock as merely one indicator of city population, among others, and eventually exclude it from the analysis. Yet it remains significant in itself, which is not the case for most other indicators of social stratification. Moreover, immigration was one of the major issues in municipal election campaigns in 1983, and these relationships probably reveal its effects. We thus choose to interpret foreign-born population as an indicator of social strain related to unemployment and economic crisis, which obviously affected the local policy arena in recent years, and to keep it in the analysis as a fundamental factor. This leads us to try to specify the nature of the effect of this important variable.

An initial explanation is strictly statistical. Retrenchment may occur in cities with large popular classes and foreign population simply because the relative level of spending is higher there. In support of this view, we see that spending per capita in 1982 is related to the rates of unemployment, size of foreign population, and the age of the population, and is limited by the level of federal grants in the city's budget (see Appendix C, Equation 3). This latter relationship reflects the relatively high level of welfare spending in cities with large low-income classes and limited property revenues, and the targeting of national grants on cities with low levels of spending. Also note that popular classes are associated with adoption of retrenchment strategies but are not significantly related to per capita spending. Since political affiliation of the mayor (either in 1977 or in 1983) also is not significantly associated with expenditures per capita, it appears the explanation of retrenchment is more structural than political.

Indeed, the major social indicators identified in the first analysis lose some of their significance when we control for relative spending in 1982 (see Appendix C, Equation 4). Population characteristics, therefore, appear to influence retrenchment strategies through their effects

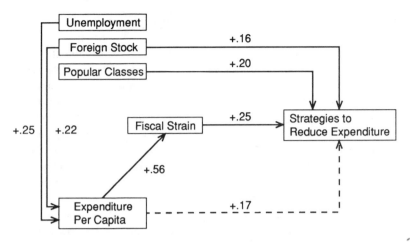

Figure 8.1 Model for Structural Determinants of Retrenchment

NOTE: The dotted line describes the hypothesized relation between spending and strategies of retrenchment, probably hidden by collinearity between spending and fiscal strain.

on levels of spending rather than through local political and social interactions. This interpretation emphasizes structural phenomena and anterior policies that shape local fiscal strategies; it portrays cities not as isolated political systems, but as part of a general politico-urban network (see Figure 8.1). It also sees city politics as an expression of macrophenomena, a view consonant with the recent structural centralization and nationalization of stakes in local elections in France. But it does not tell us how leaders have shifted their decisions.

Where Do Leaders' Preferences for Retrenchment Come From?

A number of anomalies demand further explanation. The relative importance of migrants directly affects local strategies to reduce expenditures, whatever the level of fiscal strain.[7] Furthermore, unemployment and popular classes positively affect fiscal strain while population size and foreign population tend to limit it (see Appendix C, Equation 2). The relationship of these latter variables to fiscal strain is the opposite of their relationship with strategies of retrenchment. Therefore, they cannot be considered to affect policies in the same way: Foreign population generates high levels of spending but not of fiscal strain. This indicates either that the specific effects of per capita spending cannot be captured because of its collinearity with fiscal strain or that more political explanations related to leaders' preferences are in order.

As leaders' preferences influence the selection of retrenchment strategies, we need to identify the origins of these preferences (see Appendix C, Equation 5). Overall, mayors' preferences for retrenchment strategies appear to be exogenously affected by the aging population of the city. Within the political process, they are constrained by group activity and shaped, indirectly, by political culture, especially those new populist mayors affiliated with rightist parties. Leaders in cities with growing percentages of elderly, rightist leaders and leaders reporting that they often take positions against their constituents are more likely to favor retrenchment. The constraint of greater group activity is consistent with the idea that most organized groups are fiscally liberal (generally demanding municipal intervention), and that leaders tend to be responsive to their preferences in the political process.[8]

Leaders' preferences are also shaped by the political cultures of which they are a part (see Appendix C, Equation 6). These political culture indexes, however, are significant only when they include partisan affiliation or partisan integration. Specifically, the basic influence is the indicator of rightist new populism, that is, of leaders fiscally conservative but socially liberal, more responsive to citizens than to organized groups, and affiliated with right-wing political parties. The dynamism of leaders and the conflict with organized groups in the formation of these preferences is consistent with the idea of populist leaders, while partisan affiliation shows that this phenomenon mainly occurs within rightist parties. Two points deserve mention here. First, traditional rightist leaders also exhibit these preferences, but to a lesser extent than new populists; clearly, rightist leaders are more likely to favor retrenchment when they are less socially conservative. Second, the index of leftist populism remains insignificant, showing that leftist mayors are basically opposed to retrenchment strategies.

Effects of Political Context

While these factors influence preferences, they do not directly influence policy choices. There is no direct relationship between leaders' responsiveness to citizens, their political dynamism, political culture, or partisan affiliation and the actual selection of strategies to reduce spending. These results do not contradict the previous finding that leaders' preferences do matter; rather, they specify the analysis and emphasize the differences between leaders' preference formation and effective selection of these strategies. In short, leadership matters, but is also constrained by local and political situations in the decision process.

These contextual constraints are exemplified in the analysis of the relationship of fiscal strategies and political changeovers in the last local elections (in 1983).[9] This analysis shows a positive relation between stable leftist governments and expenditure-reducing strategies; that is, the more stable the leftist municipal government, the more likely strategies to reduce expenditures have been selected. This result does not fit the general association of fiscal conservatism with rightist parties. It contradicts our expectations since retrenchment was a major issue of the right in the political campaign of 1983. It can be understood only within the historical context.

During the previous interelectoral period (1977-1983), the left dominated the government of large and middle-sized French cities. Their strong reform orientation, and lack of access to the national government before 1981 led many mayors to develop their own innovative policies, particularly for leisure, urbanism, and social welfare programs. Did this process lead to a higher level of spending in leftist cities? The answer appears to be no, as expenditure per capita in 1982 is not related to partisan affiliation of the mayor at this time, nor is fiscal strain.

If level of spending is independent of partisan affiliation, other explanations for leftist austerity must be found. A possible alternative explanation is that retrenchment is more likely to be selected by incumbent than new mayors. It is probably easier for a new mayor to increase than to decrease spending, as effects produced by high levels of taxes are delayed until the next election, while cutbacks on services can instantly generate opposition from employees and user groups. Retrenchment is a long-term process involving lengthy negotiations and procedures. Beyond the dramatic cases reported by the press, it does not proceed by massive cutbacks but by limited and gradual reductions in various areas of city government. Furthermore, identifying the type of spending to be reduced implies the active cooperation of the municipal bureaucracy and a developed knowledge of its routines. This argument suggests an interpretation in terms of organizational and bureaucratic constraints. It also implies policy cycles: New mayors are in search of political visibility; new policy outcomes are easier to exhibit through spending than through retrenchment. Political stability, therefore, favors strategies of reduction of spending in terms of both organizational and political processes.

If political stability and spending per capita are comparable, why do cities where the right remained in power in the last election not select retrenchment strategies more than others? As a strictly fiscal explanation is not adequate, consider more social factors: Both policy effects *and* political stability for leftist governments are shaped by the composition of the local population. Popular classes are related positively to the retention of power by the left in the last election,[10] thus cities that select strategies of retrenchment are more likely to have large popular classes who both favor retrenchment and ensure more stability to the left. But it is important to note that this relationship is not due to economic constraints, as popular classes are related neither to fiscal strain nor to the level of spending; the process is specifically political.

These considerations underline the gap existing between the raising of issues on political agendas and the constraints on their implementation through policy outcomes: Fiscal retrenchment was brought into the political debate through the rightist parties' campaigns of 1983 but was then more likely to be selected by both rightist mayors and leftist mayors who remained in power after the election. Leftist mayors are more frequently associated with population characteristics likely to favor retrenchment; their political stability enables them to select these strategies more easily than mayors in cities facing political change. Rather than a strict opposition between right and left models of government, this interpretation suggests an adaptation process of decisions to the issues shaping the political agenda, progressively normalizing the patterns of spending.

Leftist Austerity and Rightist Retrenchment

How does this adaptation of decisions occur? One possibility is that foreign-born populations indirectly activate the local selection of retrenchment strategies. Thus social strain (measured here through foreign-born population) may lead some groups to react to the degradation of their social conditions and threats to their status with an orientation toward fiscal conservatism. To maintain their popular class electorate, leaders in turn would select strategies to reduce expenditures and limit municipal interventions. To test this proposition, we built interaction terms between foreign-born population and other indicators for age and occupational groups. None was significant for strategies of retrenchment, or for leaders' preferences or the type of political change. So far, the lack of statistical evidence does not support such an explanation[11] and suggests that the effect of foreign-born population on policy choice is more structural than political.

A more theoretical consideration argues that policy adaptation mainly occurs through the setting of local political agendas by partisan competition. Cities with large upper-class populations favor stability of power to the right, and rightist mayors, in turn, favor the selection of strategies of retrenchment. But, as described above, stability of power to the left is associated with large popular classes, and these cities also tend to implement austerity. Both partisan political processes lead to the same policy strategies: Different social characteristics of cities produce the same outputs.

Although this indicates that retrenchment is supported by diverse social groups, the persistent relationship between retrenchment strategies and the size of the foreign population indicates that the process is far more complex than a simple change in citizen preferences toward fiscal conservatism leading, in a Downsian sense, political parties to match these preferences (Downs, 1957). Recall that the results of the 1983 election were the outcome of a situation where rightist candidates took fiscally but also socially conservative positions, especially on immigration issues, in opposition to the socialist government. The far-right party Front National led by Jean-Marie Le Pen actively campaigned on these issues and won important support in the election. Citizens may have supported these candidates because they were socially conservative and opposed to the government and only incidentally because these leaders "happened" to be also fiscally conservative. Electoral competition then would have led leftist mayors to select local fiscal strategies in conformity with their perceived evaluation of citizens' retrenchment preferences.[12]

Although this formulation is probably extreme, it recognizes that specific fiscal issues were far less important in the 1983 election than a general hostility toward the government and the immigration issue. Policy issues are interrelated and presented in packages, and our results show that they are difficult to evaluate separately. Finally, this point suggests that elections may depend more on ideological than policy issues but nevertheless may still prompt adaptation in local policy strategies. Rather than focusing on a direct and local interaction between leaders and citizens over policy issues, this perspective considers a broader level of interaction, where national partisan competition is influential in the determination of local policy outcomes (see Figure 8.2). This interpretation of local policy adaptation processes is especially appropriate for the case of France because of the relatively high level of centralization and the globalization of the stakes in local elections in recent years.[13]

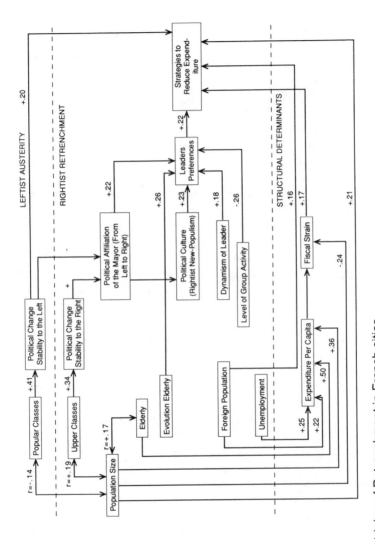

Figure 8.2 Origins of Retrenchment in French cities

NOTE: r's are Pearson coefficients; other numbers indicate beta coefficients.

Constraints on
Revenue-Raising Strategies

Strategies to increase city revenues are much more difficult to model. Indeed, they combine rather different types of policies that are not necessarily linked together. Causal patterns of specific policies could be antagonistic and their effects would be canceled when aggregated. The focus here, therefore, is on explaining strategies to raise local taxes; it does not include other revenue-raising strategies such as seeking intergovernmental grants, borrowing, or raising the level of consumer fees for services.

The analysis shows significant relationships among leaders' preferences to increase taxes, population age structure, and the level of national grants (see Appendix C, Equation 8). Cities with young populations tend to limit the use of these tax-raising strategies, suggesting a cumulative effect vis-à-vis strategies to decrease expenditures: Aging populations lead cities to limit spending, while young populations limit tax increases. The level of state participation in the budget, here considered in 1982, seems to favor the selection of strategies to raise taxes. Thus the more a city was dependent on government grants in 1982, the more the mayor called for taxpayers to support the budget through higher taxation. One of the possible effects, therefore, of decentralization reforms — national retrenchment (or at least its anticipation by mayors) — is increasing local taxes.

Leaders' preferences for raising taxes (see Appendix C, Equation 9) are far less dependent on partisan processes than are retrenchment preferences. Political variables such as partisan affiliation and integration of the mayor, type of succession in last election, and political culture are not significantly related to taxation preferences, nor are group or citizen responsiveness, the level of expenditures, or the number of municipal employees per capita. Instead, tax-increase orientations are dependent on changes in intergovernmental relations and are further constrained by local population characteristics and group activity. Specifically, the level of group activity and the percentage of foreign population tends to dampen fiscally liberal mayors' preferences for greater taxation and spending. As described above, foreign-born populations are associated with strategies to reduce spending; here, they appear also to limit strategies to increase taxes. In contrast, group activity constrains both retrenchment and taxation strategies.

Innovation and Privatization

Using the comparative index for innovative strategies,[14] innovative strategies are more likely to be adopted in cities with high proportions of middle-class groups, high levels of expenditure per capita, and direct involvement of elected officials in the budgetary process (see Appendix C, Equation 10). The importance of the budgetary decision process shows the level of politicization surrounding these types of strategies. To override the incrementalism of local bureaucracy, to implement productivity controls over employees, and to concede some services to private firms, elected officials have to be actively and deeply involved in decisions. The negative effect of the percentage of public workers probably reflects, beyond socioeconomic conditions, the importance of unions in the city's life and their expectations of local government, which limits the room for maneuver for elected officials.

This is especially evident in the case of privatization strategies. Selecting privatization strategies is related to the level of expenditure per capita, mayors' preferences, and partisan affiliation (see Appendix C, Equation 11). These preferences for privatization are deeply linked to partisan affiliation and political cultures; they are especially salient for new populist rightist mayors, who are more likely to have won the last election when a leftist mayor was previously in power. Social composition of the population does not affect these preferences although they may be constrained by relations of leaders with organized groups (see Appendix C, Equation 12). Thus the most recent generation of strategies to cope with fiscal strain is probably the one with the most exogenous origins. Unlike other innovative strategies, this strategy of privatization is not dependent on the decision process; this suggests that municipal employees and bureaucracy have much more capacity to constrain internal innovation, such as productivity within services, than other types of strategies.

Conclusion

These analyses provide some initial findings on the determinants of retrenchment, revenue, and innovative policy strategies in French cities. The findings indicate that retrenchment strategies are a general phenomenon in French cities, adopted by both rightist and leftist mayors but involving different political processes. In regard to revenue-raising strategies, the relation with central government is one of the main factors leading to increases in local taxes, since mayors are more likely to favor taxation-increase policies when their budgets are more dependent on national grants. Thus national retrenchment associated with decentralization reforms leads cities to develop their resources and increase local tax burdens. Finally, innovative strategies are more often selected by the new rightist generation of mayors and are dependent on the deeper involvement of elected officials in local decision processes. This strong leadership is associated with a partisan distribution of some of these strategies, specifically of privatization. In this we see the effects of the political campaign of 1983, where the reduction of local spending became an important national issue, and probably also of the partisan strategies of government.[15] This increased involvement of mayors and salient partisan cleavages lead us to conclude that fiscal policies in French cities have become increasingly politicized over the past few years.

Appendix A: Dependent Variables

Dependent variables are the strategies used by cities as reported by the chief administrative officers. The wording is "Please indicate if these strategies were actually implemented in your city in the last ten years." After recoding, values are "no answer" = "not applicable" = .5; "not implemented" = 1; "implemented" = 2. Categories are as follows:

- *Strategies to reduce expenditures:* LEXPSTR2 = E50 (defer payments to next year) + E51 (reduce global spending) + E52 (reduce investment spending) + E53 (eliminate programs) + E54 (reduce administrative spending) + E55 (reduce some services) + E56 (defer maintenance of capital stock) + E57 (reduce spending for maintenance of capital stock) + E58 (reduce number of municipal employees by layoffs) + E59 (reduce number of employees through retirement) + E60 (freeze hiring) + E61 (eliminate some services) + E64 (reduce subventions, or subsidies).
- *Strategies to increase revenues:* LREVSTR1 = E36 (seek new local revenues) + E37 (seek new national grants) + E38 (seek new grants from the departement + E39 (seek new grants from the region) + E40 (increase local taxes) + E41 (increase customers' fees for municipal services) + E42 (create fees for free services) + E43 (sell some assets) + E44 (long-term borrowing) + E45 (short-term borrowing) + E46 (lower "fonds de roulement") + E47 (favor a more intensive use of services to get revenues) + E48 (favor population growth) + E49 (attract new industries). Among these, we distinguished strategies to get subventions (LSUBST = E37 + E38 + E39), strategies to raise local taxes (LE40), strategies to increase

borrowing (LEMPST = E44 + E47), and strategies to increase revenues through services (LSERVST = E41 + E42 + E47).

- *Strategies to improve productivity:* We built a general index summing the same strategies as retained in the American index: LPROSTR2 = E62 (privatization) + E66 (improve productivity by control over employees) + E71 (purchasing agreements) + E72 (call for expertise in evaluation of costs and needs). We also used an index of privatization LPROSTR4 = E62 (privatization) + E65 ("debudgetization") + E70 (call for volunteers) + E71 (renegotiation of purchasing agreements).

Appendix B: Independent Variables

(1) The main indexes for *population characteristics* are as follows:

CMOY1 = M27 (managers and professionals) + M28 (intermediate professionals = middle-range employees such as nurses, social workers, and technicians)

CMOY2 = M27 (managers and professionals) + M28 (intermediate professionals) + M29 (employees = low-income white-collar workers)

CSUP = M26 (businessmen, merchants, and shopkeepers) + M27 (managers and professionals)

CPOP = M29 (employees) + M30 (workers)

CPOP2 = M28 (intermediate professionals) + M29 (employees) + M30 (workers)

We also used each of these categories separately. Values are expressed in percentages of the working population.

(2) *Fiscal strain* indicators are defined as follows:

LFS1 = L09 (total spending per capita 1982)/LO7 (potential fiscal 1982)

LFS2 = L14 (debt per capita 1982)/LO7 (potential fiscal 1982)

LFS4 = L06 (taxes per capita 1982)/LO7 (potential fiscal 1982)

(3) *Intergovernmental relations* are considered through two indicators:

LFEDGR1 = H21 (annual change in national grant for 1985)/HO6 (annual change in total spending for 1985)

This first indicator provides a measure of recent trends in central/local fiscal relations.

LFEDGR2 = K08 (national grant for 1982)/K10 (total spending for 1982)

This second index provides an evaluation of the anterior level of national grants on the development of strategies (the last election occurred in 1983).

(4) *Municipal employees sector* is measured with the following ratio:

SECPERS = I04 (number of municipal employees in 1982)/M11 (population size in 1982)

(5) *Partisan affiliation* of the mayor in 1983 (M08) and 1977 (M07) is considered as an ordinal variable; after recoding, the following continuum is created, from left to right: 1 = Parti communiste français; 2 = Parti socialiste; 3 = mouvement des radicaux de gauche; 4 = Union pour la democratie française; 5 = Rassemblement pour la Republique; 6 = divers droite. The position of divers droite on the far right is justified by the importance of the Front National in the last election. We see this classification as a first approximation, but further steps should use dummy variables: many UDF local leaders are more conservative than young RPR, for instance, and the divers droite category mixes moderate independent with far-rightist leaders.

(6) Mayors' *partisan integration* index is PARTISAN = C43 + C45 + C46; after recoding C43 ("Are you a member of a political party?"), 1 = no; 2 = regular member; 3 = party activist; 4 = local executive; 5 = national executive. C45 ("What was the role of your party during the last election campaign for municipal government in your city?"), 1 = don't know/not applicable; 2 = no activity; 3 = casual activity; 4 = active; 5 = very active. C47 ("Did you sometimes disagree with your party's activists about issues involving city government?"), 1 = very often; 2 = rather often; 3 = sometimes; 4 = seldom or never; 5 = don't know/not applicable.

(7) *Decision process* index is DECISION = A12 + A13, where A12 = "What is the respective role of elected officials and city managers in deciding the total amount of spending and its distribution among services?"; 1 = elected officials are dominant; 2 = elected officials are most important; 3 = elected officials and city managers are equivalent; 4 = city managers are most important; 5 = city managers are overwhelming; 6 = the role of elected officials and city managers is limited — the budget is the same from year to year. A13 = "What about the elaboration of the budget for each service?" (same wording for the same values).

(8) *Mayors' preferences concerning strategies* were computed as follows:
Preferences to reduce expenditures LEXPRF2 = B04 (reduce investment spending) + B05 (reduce capital spending) + B06 (stop growth of the number of municipal employees) + B07 (freeze salaries) + B13 (reduce some services).
Preferences to increase revenues LREVPRF2 = B08 (increase fees for services) + B09 (raise local taxes) + B10 (seek new national grants) + B11 (seek

other intergovernmental grants) + B12 (borrowing). Adjustments were made to fit the subcategories distinguished above: LSUBPRF = preferences to seek new intergovernmental grants = B10 + B11; preferences to raise taxes = LB09; preferences to develop borrowing = LB12; preferences to get revenues through services = LB08.

Preferences to improve productivity are generally not included in the questionnaire; we nevertheless used B14 (preferences for privatization of municipal services) per se and as a general indicator for innovation. However, it showed no significance when considered in this manner.

(9) *Fiscal liberalism of leaders* is INTERFI1 = A24 + A27, where A24 = "In your opinion, should local taxes be reduced in your city?" (from 1 = much reduced to 5 = much raised). A27 = "For all areas of city government, would you like to spend . . . ?" (from 1 = far less to 5 = far more).

(10) *Social conservatism of leaders* is CSCULT2 = C51 + C52, where C51 = "What is your personal position about legalization of abortion?" C52 = "What is your personal position about the death penalty?" We recoded to create a progressive scale: 1 = liberal; 5 = don't know; 2 = conservative. We did not use a third question regarding stricter control of immigration because there is great consensus among mayors and, consequently, this issue does not differentiate among them.

(11) *Group responsiveness* is computed as GRRESP2 = C04 (municipal employees' organizations) + C05 (users of public services' organizations) + C06 (neighborhood groups) + C07 (environment and ecologist movements) + C08 (tenants' unions) + C09 (homeowners' groups) + C10 (employees' unions) + C11 (chambers of commerce) + C12 (business unions) + C13 (local associations of shopkeepers) + C14 (leisure groups) + C15 (taxpayers' associations) + C16 (elderly groups) + C17 (youth movements) + C19 (city councils) + C20 (city managers). Also, mayors were asked to evaluate the *level of activity* of these groups; ACT1 is computed as the sum of scores for each of them.

(12) Political culture indexes are as follows:

DROITE = 0 − INTERFI2 + CSCULT3 − GRRESP22 + C18
GAUCHE = 0 + INTERFI2 − CSCULT3 + GRRESP22 − C18
NFP = 0 − INTERFI2 − CSCULT3 − GRRESP22 + C18
DROITE1 = DROITE + PARTISA2
DROITE2 = DROITE − PARTISA2
GAUCHE1 = GAUCHE + PARTISA2
GAUCHE2 = GAUCHE − PARTISA2
NFP1 = NFP + PARTISA2
NFP2 = NFP − PARTISA2
NFP3 = NFP + M08 (rightist new populists)
NFP4 = NFP − M08 (leftist new populists)

where INTERFI2 = fiscal liberalism; CSCULT3 = social conservatism; GRRESP22 = group responsiveness; C18 = citizen responsiveness; PARTISA2 = partisan integration of the leader; and M08 = political affiliation of mayor in 1983. Notice that each of the basic dimensions of political cultures is included in the index by dividing the score by the number of variables retained in the index; for example, CSCULT3 = (C51 + C52)/2, GRRESP22 = GRRESP2/16. One of the assumptions of this definition of political culture is that each of these dimensions is equally important.

As they are intercorrelated, these political culture indexes were included one by one in each of our regressions.

Appendix C

Regression Results

Numbers indicate beta coefficients; numbers in parentheses indicate level of significance for each variable. R^2s are not adjusted.

$$LEXPSTR2 = K + .220LEXPRF2 + .219LM11 + .185LM18 + .172LFS2$$
$$\qquad\qquad (.023) \qquad\quad (.023) \qquad\quad (.057) \qquad\quad (.076)$$
$$R^2 = .156; \ i = 101; \qquad\qquad SigF = .0054 \qquad\qquad\qquad\qquad [1]$$

where LEXPSTR2 = strategies to reduce expenditures; LFS2 = fiscal strain as a function of debt; LM11 = population size; LM18 = change in foreign population between the last two censuses; LEXPRF2 = leaders' preferences concerning strategies to reduce expenditures.

$$LFS2 = K + .145LFEDGR2 + .561LL09 - .075M15 - .248M17$$
$$\qquad\qquad (.139) \qquad\qquad (.000) \qquad\quad (.398) \qquad\quad (.000)$$
$$\qquad + .152M20 - .222M27 - .246LM11$$
$$\qquad\quad (.058) \qquad\quad (.015) \qquad\quad (.018)$$
$$R^2 = .339 \ \ i = 168 \qquad SigF = .0 \qquad\qquad\qquad\qquad [2]$$

where LFS2 = level of fiscal strain; LL09 = expenditures per capita; LFEDGR2 = percentage of national grants in the budget; M15 = rate of unemployment; M17 = percentage of foreign stock; M20 = percentage of population younger than 15; M27 = managers and professionals; LM11 = population size.

$$LL09 = K + .032M07 - .188LFEDGR2 + .257M15 + .073LM16$$
$$\qquad\qquad (.651) \qquad (.009) \qquad\qquad (.011) \qquad (.489)$$
$$\qquad + .225M17 + .178LM18 - .504M20$$
$$\qquad\quad (.004) \qquad (.033) \qquad\quad (.000)$$
$$R^2 = .255 \ \ i = 169 \qquad SigF = .0 \qquad\qquad\qquad\qquad [3]$$

where LL09 = spending per capita in 1982; M07 = partisan affiliation of the mayor in 1977; LFEDGR2 = level of national grants in the city's budget in 1982; M15 = percentage of unemployment; LM16 = change in rate of unemployment; M17 = foreign stock; LM18 = evolution of foreign stock; M20 = percentage of population younger than 15.

$$LEXPSTR2 = K - .010CPOP2 + .175LL09 + .234LEXPRF2 + .111M17$$
$$(.933) \quad\quad (.096) \quad\quad (.018) \quad\quad\quad (.308)$$
$$+ .160M15 - .247M26 - .053M22$$
$$(.174) \quad\quad (.110) \quad\quad (.728)$$
$$R^2 = .164 \quad i = 101 \quad\quad SigF = .015 \quad\quad\quad\quad [4]$$

where LEXPSTR2 = strategies to reduce expenditure; CPOP2 = popular classes; LL09 = spending per capita 1982; LEXPRF2 = leaders' preferences to reduce expenditures; M15 = level of unemployment; M17 = foreign stock; M22 = percentage of elderly; M26 = traditional upper class.

$$LEXPRF2 = K + .220M08 + .181C47 - .261ACT1 + .264LM25$$
$$(.010) \quad\quad (.033) \quad\quad (.003) \quad\quad (.002)$$
$$R^2 = .262 \quad i = 108 \quad\quad SigF = .0000 \quad\quad\quad\quad [5]$$

where LEXPRF2 = leaders' preferences for strategies to reduce expenditures; M08 = political affiliation of the mayor from left to right; C47 = political dynamism of the mayor; ACT1 = level of activity of organized groups; LM25 = change in percentage of elderly.

$$LEXPPRF2 = K + .231NFPD3 + .260LM25 - .270ACT1$$
$$(.013) \quad\quad (.006) \quad\quad (.005)$$
$$R^2 = .229 \quad i = 94 \quad\quad SigF = .005 \quad\quad\quad\quad [6]$$

where NFPD3 = rightist, new populist political culture; LM25 = evolution in percentage of elderly; ACT1 = level of activity of all organized groups.

$$LEXPSTR2 = K + .286LEXPRF2 + .259LFS2 + .236LM11 + .205GG$$
$$(.003) \quad\quad\quad (.007) \quad\quad (.011) \quad\quad (.086)$$
$$+ .161M17 - .210M26 + .072DD$$
$$(.093) \quad\quad (.039) \quad\quad (.541)$$
$$R^2 = .236 \quad i = 101 \quad\quad SigF = .0005 \quad\quad\quad\quad [7]$$

where LEXPSTR2 = strategies to reduce expenditures; LFS2 = fiscal strain; LM11 = population size; M17 = foreign stock; M26 = businessmen, merchants, and shopkeepers; GG = stability of mayor to the left in last election; DD = stability of mayor to the right.

$$LE40 = K + .219LB09 - .252M20 + .303LFEDGR2 - .182LM11$$
$$(.022) \qquad (.017) \qquad (.023) \qquad\qquad (.177)$$
$$+ .125GG + .040GD$$
$$(.238) \qquad (.701)$$
$$R^2 = .181 \quad i = 101 \qquad SigF = .003 \tag{8}$$

where LE40 = strategy to increase taxes; LB09 = leaders' preferences to increase taxes; M20 = percentage of city population younger than 15; LFEDGR2 = level of federal grants; LM11 = population size; GG = stability of mayor to the left; GD = change of mayor from the left to the right in the last election.

$$LB09 = K + .249INTERFI1 - .187M17 - .257ACT1 + .037LL09$$
$$(.006) \qquad\qquad (.031) \qquad (.003) \qquad (.677)$$
$$R^2 = .190 \quad i = 118 \qquad SigF = .000 \tag{9}$$

where LB09 = leaders' preferences to raise taxes; INTERFI1 = fiscal liberalism of leaders; M17 = foreign stock; ACT1 = level of activity for all groups; LL09 = level of expenditure per capita 1982.

$$LPROSTR2 = K + .280CMOY1 + .213LL09 - .182DECISION$$
$$(.005) \qquad\quad (.026) \qquad (.004)$$
$$R^2 = .136 \quad i = 101 \qquad SigF = .002 \tag{10}$$

where LPROSTR2 = strategies to improve productivity; CMOY1 = upper-middle class; LL09 = expenditure per capita 1982; DECISION = importance of managers and services compared to elected in the elaboration of the budget.

$$LE62 = K + .197LB14 + .217LL09 + .189M08 - .125LM16 + E$$
$$(.050) \qquad (.020) \qquad (.059) \qquad (.187)$$
$$R^2 = .179 \quad i = 101 \qquad SigF = .000 \tag{11}$$

where LE62 = strategy of privatization; LB14 = mayor's preference for privatization; LL09 = level of expenditures per capita 1982; M08 = political affiliation from left to right; LM16 = evolution of unemployment.

$$LB14 = K + .356GD + .024GG + .252NFPD3 - .231ACT1 - .029M20$$
$$(.001) \qquad (.856) \qquad (.086) \qquad (.013) \qquad (.762)$$
$$- .063LFEDGR2$$
$$(.501)$$
$$R^2 = .309 \quad N = 93 \qquad SigF = .000 \tag{12}$$

where LB14 = leaders' preferences for strategy of privatization; GD = type of succession in last election (from left to right); NFPD3 = rightist, new populist political culture; ACT1 = level of activity of all organized groups; M20 = percentage of population younger than 15.

Notes

This chapter was written by Richard Balme while he was a visiting scholar at the University of Chicago. The data were collected in France by Jeanne Becquart-Leclercq, Jean-Yves Nevers, and Vincent Hoffmann-Martinot. The author also acknowledges the advice of Terry N. Clark and Susan Clarke in conducting the analysis and editing the final text.

1. Correlations between dependent variables vary between .34 and .38.

2. Foreign population, as counted by the French census, includes only long-term residents without French citizenship. Therefore, this category is not equivalent to foreign stock, as counted by the American census, because this also includes American citizens.

3. The fiscal potential is a measure where the tax base, including property value, is weighted by its average for cities of the same size.

4. Each of these measures of fiscal strain was alternatively used in each of our regressions.

5. As this measurement is based on the mayors' assessment, it is clearly subjective and should later be supplemented by counterevaluations made by city managers.

6. Foreign population size is correlated positively with large popular classes (r = .227) and negatively with upper-class (−.147) and elderly (−.354) indicators.

7. The effects of fiscal strain and level of spending are difficult to evaluate jointly as they are correlated (r = .412). When we consider them together, fiscal strain overrides the level of spending, which shows no effect in the equation. Moreover, fiscal strain is determined by the level of spending and, therefore, already takes into account its effect (see Equation 2, Appendix C). On the other hand, the percentage of foreign population remains significant in the regression when we control for the level of fiscal strain, suggesting other types of explanations.

8. Group activity and group responsiveness of leaders are correlated and have similar effects when included in the equation.

9. This dimension was used both as an ordinal variable ranking stability from the left to stability to the right (1 = left to left with n = 73; 2 = right to left with n = 4; 3 = left to right with n = 33; 4 = right to right with n = 61), and as a dummy variable, creating one variable for each of these types. Both methods showed consistent results. See Appendix C, Equation 7.

10. r = .414 between CPOP2 (large popular classes) and GG (stability to the left).

11. We found an interaction effect between elderly and foreign population on stability of power to the right; this disappears, however, when we control for upper classes.

12. This proposition is supported by the fact that leaders' preferences to reduce spending are determined by their affiliation in rightist parties. Leftist mayors are thus oriented toward retrenchment primarily through electoral competition.

13. In 1983, 81% of French citizens knew the names of their mayors, in comparison to 65% in 1975. Some 73% thought that these elections are mainly political (as opposed to local) for the whole of the country; 50% vote for the personality of candidates, and 43% for their partisan affiliation. For cities over 30,000 (roughly corresponding to our sample), 54% of voters determine their choice through partisan affiliation, and 40% through personality of candidates (SOFRES, 1984).

14. LPROSTR2 is a sum of four innovative strategies; see Appendix A.

15. After the municipal elections of 1983, several rightist mayors experimented with privatization. Jacques Chirac, mayor of Paris, was a pioneer in this domain; he became

prime minister in 1986 and Edouard Balladur, his main assistant in city hall, became Minister of Finances and Privatization.

References

Clark, T. N., Burg, M. H., & de Landa, M.D.V. (1984). *Urban political cultures and fiscal austerity strategies.* Paper presented at the annual meeting of the American Political Science Association, Washington, DC.

Clark, T. N., & Ferguson, L. C. (1983). *City money: Political processes, fiscal strain, and retrenchment.* New York: Columbia University Press.

Downs, A. (1957). *An economic theory of democracy.* New York: Harper & Row.

Levy, F. S., Melstner, A. J., & Wildavsky, A. (1974). *Urban outcomes.* Berkeley: University of California Press.

Lowi, T. (1969). *The end of liberalism.* New York: Norton.

Peterson, P. E. (1981). *City limits.* Chicago: University of Chicago Press.

Shefter, M. (1985). *Political crisis, fiscal crisis.* New York: Basic Books.

SOFRES. (1984). *Opinion publique 1984.* Paris: Gallimard.

Wildavsky, A. (1964). *The politics of the budgetary process.* Boston: Little, Brown.

9

The Political Implications
of Fiscal Policy Changes:
An Overview

SUSAN E. CLARKE

This volume uses the concepts of autonomy and innovation as the lens through which to view local fiscal policy change in Europe and the United States. These concepts guided the individual country analyses and shaped the interpretations of local policy change; they are used here to provide a more structured interpretation of these research findings as a whole. The concept of local autonomy informing this volume draws on Clark's (1984) identification of two dimensions of state institutional autonomy: local governments' ability to initiate actions to achieve their goals and their immunity from oversight by higher state authorities. Gurr and King's (1987) differential autonomy argument expands on this approach. They distinguish between Type I autonomy, the local state's ability to pursue its interests without constraints of local social and economic forces, and Type II autonomy, the ability to act without interference from other segments of state authority.

These concepts center on a key feature of contemporary local politics: Local officials must deal with an environment largely shaped by economic and political forces beyond their control. But the local nexus of these forces creates pressing concerns. There is an "imperative" (Peterson, 1981) to accommodate business interests in order to generate sufficient local revenues. With the exception of the Netherlands, local officials in the communities studied (and in most Western democracies) primarily rely on local sources for the majority of their revenues. But at the same time, they face electoral imperatives to respond to the needs and demands of local citizens. Local officials, therefore, must devise strategies that protect their standing with these

236

disparate groups while expanding their discretionary "political space" to manage these often dissonant pressures.

Further constraints stem from the context of central-local relations. As distinctive entities in a fragmented state apparatus, the interests of local governments may conflict with certain central state objectives (Gurr & King, 1987, p. 70). This is particularly so in the fiscal policy arena. Central governments often resort to manipulating central-local linkages in order to moderate local governments' "claimant" interests on resources and to protect broader economic objectives (Schick, 1988, p. 524). In current efforts to cope with economic restructuring, this manipulation can be especially severe and consequential. Local officials, therefore, must seek strategies that not only balance tensions within the community but also address these interests in managing conflict between the local community and central government. To do anything less jeopardizes their ability to maintain and sustain the authority and legitimacy of governance institutions.

The analyses of such efforts in the preceding country studies provide an unusual opportunity to examine systematically local political change in the 1980s. This concluding chapter draws out some of the themes and trends characterizing this complex relationship of local political autonomy and fiscal policy innovation.

Central-Local Relations and Fiscal Policy Change

The complexity of tax codes, borrowing regulations, and grant-in-aid structures makes comparative analyses of local finance troublesome even within the same country. Comparative cross-national analyses of central-local fiscal relations face additional problems of equivalence and standardization. The FAUI Project circumvents these problems by coordinating longitudinal fiscal data archives provided by researchers in each country according to standardized procedures (Mouritzen & Nielsen, 1988) with systematic collection of survey data on local officials' fiscal management strategies. The following observations on the relationship of central-local relations with fiscal stress and distinctive patterns of fiscal policy innovations are based on data from these two sources.

Local resource flexibility is increasing over time. Resource flexibility, seen here as the type and diversity of revenue generation options available to local officials, is a gauge of the actual strength of local

TABLE 9.1

Changes in Local Fiscal Discretion: Municipal Revenues per Capita, 1985
(1978 = 100)

	General Grants	Conditional Grants	Total Tax Effort	Fees and Charges
Netherlands	NA	NA	118.2	143.8
France	132.1[a]	124.6[a]	119.0[a]	123.5[a]
Norway	159.9	145.8	93.9	188.5
United States	82.0	81.4	95.9	124.7
Denmark	48.7	114.5	114.8	121.4
Finland	NA	NA	91.9	115.9[a]

SOURCE: Mouritzen and Nielsen (1988, pp. 27, 28, 30).
a. As of 1984.

government relative to institutions in both the public and private sector (Ashford, 1979, p. 80). Thus resource flexibility is an indirect measure of relative autonomy: Communities with more diverse and more flexible revenue options are considered to have greater fiscal discretion to take initiatives relatively free from oversight.

Table 9.1, showing per capita measures (in constant 1978 dollars) of changes in a range of discretionary revenues, including dependence on general rather than conditional grants, taxes, and use of fees and charges, suggests a trend toward greater revenue diversification. This diversification and increased local revenue discretion is most apparent in France and Norway; in both countries, the per capita growth in less restrictive general grants is stronger than growth in conditional grants (although French grants began to decline in 1984), there is greater use of fees and charges, and, in France especially, greater efforts to extract revenues from the local tax base.

This trend toward revenue diversification increases the complexity of local revenue structures (Sharp & Elkin, 1987). It affords greater flexibility and elasticity to local officials, with an implicit increase in autonomy, but may reduce the accountability of local fiscal policymakers, particularly if these new sources are relatively invisible (Sharp & Elkin, 1987, p. 391).

There is no consistent relationship between central-local relations and fiscal stress. We assumed, not without reservations, that the degree of fiscal centralization in central-local relations roughly indicates the local discretion and initiative available to local officials. These six countries vary in their reliance on central government grants and local

TABLE 9.2

Fiscal Stress Measures

	FAUI Fiscal Stress Index 1979-1982 (less than 100 = fiscal stress)	FAUI Fiscal Stress Index 1982-1984	Debt per Capita 1984[a]	Tax Effort Index 1984[a]
Netherlands	119.2	103.1	NA	117.5
France	107.6	108.0	97.1	119.0
Norway	111.0	107.5	89.0	94.0
United States	95.4	101.1	113.6	96.3
Denmark	106.0	98.0	62.5	113.2
Finland	109.3	114.7	103.2	92.0

SOURCE: Mouritzen and Nielsen (1988, p. 39).
NOTE: This index operationalizes Wolman and Davis's (1980) definition of fiscal stress: It measures how much revenue (local taxes and grants) a city will require in the base year to provide the same level of services as two years previously. It takes into account increased revenue needs due to changes in needs (generally, measured as population change) and inflation. An index value of 100 indicates that expenditure needs can be met with existing tax effort. A value below 100 indicates fiscal stress, or the need either to increase tax effort to maintain services or to reduce service levels (or a combination), while a value above 100 indicates fiscal slack. The percentage columns in this table use 1979 and 1982 as the base years. See Mouritzen and Nielsen (1988, pp. 222-223) for computations used.
a. In constant dollars; 1978 = 100.

taxes for revenue, but, aside from the Netherlands, differences in fiscal dependence are not great: Each generates more than two-thirds of its revenue from the local economy. The initial expectation is that local governments depending significantly on central grants are less autonomous but more protected from the exigencies of fiscal stress by central administrations.[1] Table 9.2 reveals that, indeed, local fiscal stress in the decentralized American federal system was greater than in the more centralized European political systems from 1979 to 1982.[2] And local governments in the Netherlands, with the most restricted fiscal autonomy, were most immune from fiscal stress during this period. But apart from these gross distinctions, the relationship between greater central control and less fiscal stress is not consistent. This is the case whether measuring fiscal stress as increased tax effort or by the composite fiscal stress index. Differences in fiscal stress in the Danish, Finnish, Norwegian, and French systems cannot be traced directly to variations in fiscal dependence.

With the exception of the United States, local officials place importance on expenditure reduction as a fiscal management strategy. While increased productivity efforts were ubiquitous, many local officials placed greatest importance on, and claimed using, some of the most

TABLE 9.3

Strategies Ranked as Most Important: Index Scores

Netherlands	lay off personnel (62)	make cuts in all departments (55)	encourage early retirement (49)
France	freeze hiring (70)	reduce administrative costs (67) increase taxes (67)	lower surplus (66) increase management productivity (66)
Norway	make cuts in some departments (67)	increase user fees (60)	increase long-term debt (45) increase labor productivity (45)
United States	increase user fees (58)	add local revenues (54) increase management productivity (54)	decrease work force through attrition (49)
Denmark	reduce capital expenditure (81)	increase labor productivity (72) increase taxes (72)	increase management productivity (71) lower surplus (71)
Finland	make cuts in all departments (77)	increase management productivity (67) lower surplus (67)	make cuts in least efficient departments (65)

SOURCE: Calculated from Mouritzen and Nielsen (1988, p. 13).
NOTE: The top three strategies reported by each country are presented. Mouritzen and Nielsen recalculated strategy scores to facilitate comparisons, so these index scores do not necessarily correspond to those reported in individual chapters. See note 3, this chapter, for description of their calculations.

visible and difficult expenditure options — cuts in selective programs, layoffs, reductions in capital expenditures. But, according to Table 9.3, local officials in the United States are more inclined to respond to fiscal austerity by enhancing their discretionary revenues than by entertaining the expenditure strategies used by European local officials.[3] In contrast, with the exception of the Netherlands, where this is a less feasible option, local officials in the other European countries ranked revenue-generation strategies second to specific expenditure policies.

Local officials in more centralized systems are more likely to make "hard choices" on expenditure strategies. The top three expenditure strategies for each country are shown in bold print in Table 9.4. The listing of expenditure strategies roughly ranges from those with highly visible, concentrated costs such as selectively cutting department budgets to those that postpone choices, such as deferring maintenance, or have less visible, less concentrated costs. Judging from Table 9.4, cities in more centralized central-local systems are more likely to make

TABLE 9.4

Most Important Expenditure Strategies: Index Scores

	Netherlands	France	Norway	United States	Denmark	Finland
Make cuts in least efficient departments	NA	38	67	25	NA	65
Eliminate programs	15	9	10	26	44	34
Make cuts in all departments	55	44	14	38	25	77
Decrease services from own source revenues	NA	56	14	23	44	27
Decrease services from IGR revenues	NA	NA	9	22	36	13
Lay off personnel	62	4	4	30	26	30
Freeze wages	NA	31	41	22	59	59
Reduce capital expenditures	47	65	27	39	81	64
Reduce administrative expenses	NA	67	15	32	35	60
Defer maintenance	NA	32	41	22	59	39
Freeze hiring	NA	70	22	41	51	40
Encourage early retirement	49	NA	2	3	NA	22
Lower surpluses	42	66	39	44	71	67
Increase labor productivity	NA	31	45	45	72	64
Increase management productivity	NA	66	42	54	71	67

SOURCE: Mouritzen and Nielsen (1988, p. 13).
NOTE: As noted in the text, these strategies are ranked roughly according to political and administrative costs of making these choices. Mouritzen and Nielsen recalculated strategy scores to facilitate comparisons, so these index scores do not necessarily correspond to those reported in individual chapters. See note 3 for description of their calculations.

the hard choices, either because they are compelled to by central government expenditure controls or are sheltered from the costs of doing so by reliance on central government revenues. The lack of complete data on the Dutch case hampers any firm conclusions, but the former argument seems more compelling. Local Dutch officials make harder choices but their ratings for across-the-board expenditure cuts and cuts in capital expenditures reflect central government reinstatement of spending controls. Similarly, Finnish officials' willingness to cut expenditures can be traced to the periodic economic agreements on

TABLE 9.5

Strategy Dissimilarity: Greatest Differences in
Country Index Score and Mean Index Score[a]

	Netherlands	France	Norway	United States	Denmark	Finland
Reduce capital expenditure (54)			−27		+27	
Make cuts in least efficient departments (49)				−24		
Increase taxes (47)					+25	
Freeze hiring (45)		+25				
Reduce administrative costs (42)		+25	−27			
Make cuts in all departments (42)			−28			+35
Freeze wages (42)				−20		
Increase IGR aid (35)[b]					−27	+28
Add local revenues (34)	−19			+20	−26	+29
Cut services from own funds (33)[b]		+23				
Make purchasing agreements (32)[b]		+24	−29			
Lay off personnel (26)	+36					
Encourage early retirement (22)[c]	+27					
Defer payments (20)				−22		

a. The mean index score for these six countries is in parentheses next to each strategy listed. Recalculations are based on data in Mouritzen and Nielsen (1988, p. 13).
b. Mean excludes Netherlands (data not available).
c. Mean excludes Denmark and France (data not available).

general levels of spending negotiated between national associations of municipalities and national officials. Finnish national government retained strong financial controls over local spending even during growth periods (Schick, 1988, p. 525); this negotiation process institutionalizes efforts to harmonize central and local fiscal interests. Several other countries report that national administrations sought such fiscal constraints; no other country studied here actually implemented these direct local fiscal controls.

Distinctive local fiscal policy profiles are evident across countries. While the strength of the index scores gives some indication of the strategy importance within each country, Table 9.5 highlights strategy dissimilarities across countries. It lists the three strategies for each country that show the greatest differences from the overall mean index score for all six countries. The sharpest differences are over whether or not to seek new local revenues; local officials in the Netherlands and Denmark were much less likely to rank this as an important option. French local officials put a much stronger emphasis on managerial strategies such as reducing administrative costs and coordinating purchasing agreements than local officials in the other countries; Norwegian officials were much less likely to do so.

Political Context and Fiscal Policy Changes

The absence of big innovations, the prevalence of these piecemeal adaptive policy responses, is not peculiar to these countries or this particular study. Schick (1988, p. 523) attributes similar trends in OECD countries to factors that are germane to this study: the extent to which officials may see budgeting and financial management procedures as less relevant to their fiscal problems than the need for political will, the shortened time perspectives that encourage tactical rather than structural reforms, and the growing insularity and shrinking boundaries of budgetary processes. In each instance, the implication is that hard choices will be made through political processes rather than administrative procedures. Thus we now turn our attention to the factors that mediate the effects of central government constraints and shape local officials' fiscal policy choices — but we do so without a good deal of theoretical guidance. As Schick (1983) points out, the political processes surrounding decremental budgeting practices are likely to be quite distinct from the more familiar politics of incremental budgeting.

Local administrative networks shape fiscal policy strategies. The existence of a large, professional, local administrative structure is a significant factor shaping local fiscal policy responses in the unitary systems such as France, Denmark, and Norway. They represent the local presence of a national welfare state and are the means by which local officials are vertically integrated with national agencies. There are two views of their policy effects: in terms of direct political action and potential sources of resistance to policies that would threaten their interests as employees and as indirectly signaling the presence of client groups who could potentially be mobilized to defend proposed expenditure cuts. The results here do not conclusively support either argument. Appleton and Clark find some tentative, counterintuitive evidence that more powerful administrative staffs are more likely to adopt productivity measures, perhaps reflecting changing professional norms, but Balme finds that such measures are more likely to be adopted in French cities where staff power is weak. Mouritzen, on the other hand, sees extensive networks of public employees and clients in local communities as successfully defending the welfare state in Denmark; policy areas not penetrated by the national welfare state and not organized by these networks were most likely to experience budget cuts.

The role of political parties in fiscal policymaking is problematic. This is true for two reasons: the inconsistent relationships between partisan affiliation and policy preferences and the utility of national party structures in articulating and aggregating citizen preferences on local fiscal issues. In most communities, the political interests and technical expertise to deal with fiscal austerity initially resided with local bureaucrats. Yet the expectation was that elected officials' political affiliations would encourage certain preferences — that liberal parties would avoid expenditure cuts while conservative parties would seek them out. In Norway, the United States, France, and Denmark, there is some evidence of partisan effects on policy choices, but not always in the direction anticipated. In the United States and France, in particular, factors such as political culture and racial concerns moderate the usual effects of partisan affiliation.

Balme's description of "leftist austerity" is compelling. He argues that social and economic tensions within traditional leftist constituencies are not fully and accurately conveyed through electoral channels. As a consequence, many incumbent leftist mayors interpreted local votes informed by conservative social and racial concerns as indicating fiscal conservatism as well, and endorsed fiscally conservative programs in response. This difficulty in disentangling fiscal preferences from other electoral issues diminishes the utility of political party structures, particularly if parties are organized on class bases that do not fully reflect emergent complex values. It also is affected by splits between national and local party organizations over fiscal retrenchment strategies. This is especially striking in the Netherlands; the Dutch authors recognize distinct national party positions on fiscal retrenchment but argue that there is, thus far, a strong bipartisan local political consensus on the need to protect the social welfare of disadvantaged groups. This trend toward ambivalent ideological and partisan stances on fiscal policies and a reversal of national-local welfare orientations is an unexpected and pervasive development.

Emergent new political cultures characterize local fiscal policy. One reason traditional political parties appear less effective in local fiscal policymaking is the complicating factor of political cultures. Appleton and Clark make this argument most vigorously. They point out that the most distinctive fiscal policy profiles emerge in cities with Democratic and Ethnic political cultures. Clark, Balme, and Miranda (1988) go on to argue that when these political cultures are associated with social heterogeneity, low-income and minority populations, and active organized interest groups, they create a political context that enables "new mayors" such as Peter Flaherty in Pittsburgh and Alain Carignon in Grenoble to take dynamic leadership roles as new fiscal populists. There is modest statistical support but compelling substantive arguments for further comparative analysis of the sources, dimensions, and effects of fiscal populism. Both European and American local officials increasingly describe themselves in terms of this combination of social liberalism and fiscal conservatism (Balme, Becquart-Leclercq, Clark, Hoffmann-Martinot, & Nevers, 1987; Clark et al., 1988).

National associations of local officials can broker local-national economic policy. The extent to which associations of local officials bargain with national officials over fiscal policy is an important determinant of the discretion ultimately available to local officials. These associations were important arbitrators of formal central-local relations in the Netherlands, France, Norway, Denmark, and Finland. In the last two countries these associations negotiate economic agreements with the national governments to coordinate macroeconomic policy goals and local government policies. These agreements are a means to coordinate the different objectives of policymakers involved in macrobudgeting and microbudgeting processes; they overcome the tensions between the goals of "claimants" and "conservers" at different government levels that otherwise hamper meaningful budgetary change (Schick, 1988, p. 524). In the other countries the negotiations are less formal but similarly effective in accommodating these needs. For example, the Association of French Mayors and the Union of Dutch Municipalities act as advocates for local governments and have been able to curb national efforts to control local expenditures through extensive lobbying efforts. In the French case, this brokering role is complemented and sometimes overshadowed by *cumul des mandats* (Nevers, 1988), the presence of local officials in national governing bodies, although these clientelistic relations appear to be breaking down in many countries. In the Netherlands, the informal power of the Union of Dutch Municipalities is enhanced by its role in appointing the head of the Council of Municipal Finances, the government's advisory group on municipal financial affairs. These associations appear especially important in forestalling real local expenditure controls; instead, they promote benign "soft austerity" policies that generally protect the discretion of local officials while encouraging cooperation with recommended ceilings on local expenditures.

Even buttressed by these powerful associations, however, local officials in the countries studied appeared to have lost some of their bargaining power with national officials and to have become more vulnerable to economic change. In the Netherlands, France, and Denmark, this brokering role has been sundered by national recessions or electoral change. In France, conservative party victories in local elections prompted the socialist government to take a hard line on growth in local transfer payments and to further devolve expenditure responsibilities. The national response to fiscal austerity was even more direct in the Netherlands: General funds to local governments were cut, local expenditure cuts were mandated, and, in 1984, local governments were forced to bear a larger share of the costs of unemployment benefits.

Political decentralization strategies mask devolution of austerity. Efforts toward decentralization of some taxing authority, grant management, and expenditure responsibilities (including privatization strategies) are under way in each of the countries studied here, with the exception of the Netherlands. These decentralization policies predate fiscal austerity in many cases. The French, for example, trace decentralization pressures to demands for greater local autonomy on the part of newly elected leftist local officials in the 1960s and early 1970s. The acceleration of decentralization efforts with the incoming Mitterrand administration in the early 1980s primarily affected departmental and regional administration; the local effects, at the commune level, were more modest. They included, however, greater local control over commune budgets as well as greater flexibility and freedom in terms of taxation authority and institutional relations at the local level. Yet these new freedoms initially obscured what the French now describe as the decentralization of austerity. These apparent gains in local autonomy, even with continued revenue support from central government, mean that French local governments are now intimately involved in the management of national fiscal crisis and have lost some of their original bargaining power with national officials. In turn, they are more vulnerable to fluctuations in the national economy and less independent from local interests.

Trends in the Political Effects
of Fiscal Policy Changes

Whether or not these local strategies actually "manage" local fiscal stress or alleviate local fiscal crises, they do change local political dynamics and central-local relations in important ways. Among the more significant political effects of these local fiscal policy changes noted by FAUI researchers are the greater potential for local conflict stemming from political decentralization strategies, the increasing influence of business and taxpayer interests, the emergence of entrepreneurial local policy climates, and qualitative changes in local democratic practices.

The Two Faces of Decentralization

In the countries analyzed here, national decentralization strategies grant local officials more autonomy over fewer resources. As Hoffmann-Martinot and Nevers so vividly argue, this intensifies pressures on local officials and increases the potential for local political conflict. In a context of slowing national growth, conflict increases between levels of government as well as between elected local officials attempting to control local budgets and local administrators of centrally and locally funded programs (see Levine & Posner, 1981). Increased revenue diversification is a frequent ameliorative response. National policies increasingly allow local governments to maintain or increase expenditures by joint funding with other governments or private sector partners, to tax new revenue sources, or to increase local borrowing. Some of these new off-budget institutional arrangements may have long-term effects on the structural relations of central-local governments; although they increase local discretion over spending in some areas, in the long run, these local policy initiatives may well prove counter to national macroeconomic policy goals.

Increasing Influence of Business
and Taxpayer Interests

Many theorists anticipate that cities faced with fiscal austerity will cater to business preferences for lower social expenditures and socialization of costs that support business development (e.g., Fainstein & Fainstein, 1983; Piven & Friedland, 1984). To date, empirical studies support this predicted increase in influence of revenue providers in

both European and American communities, independent of central-
local relations (e.g., Przeworski, 1986), but fail to confirm the social
expenditure implications (e.g., Rubin & Rubin, 1986; Wong, 1985). In
these FAUI studies, greater business influence is notable in France, the
Netherlands, Norway, and the United States despite differences in local
autonomy in these countries. These trends are starkest in the context of
Dutch centralization. In the Netherlands, the historically weak autonomy of local officials in the face of a financially and administratively
strong central government is now complemented by expanded relations
with the private sector, spurred on by the limited revenue options
available to Dutch local governments. Kreukels and Spit see a consequent reemphasis on economic and technological interests and some
indication of a slighting of social welfare concerns and organized
groups. Less organized but vocal taxpayers' groups demanding greater
efficiency and prudent local fiscal policies complement this business
orientation. Mayors and administrators in each of these countries increasingly heed this volatile constituency, even when it means turning
away from traditional supporters. Yet it is unlikely that fiscal issues can
unite either taxpayers or businesses as a class because the diverse
members of these groups do not share coherent interests in a broad set
of fiscal strategies. Nevertheless, new political cultures and the break-
down of traditional political structures such as machines and clientelist
links compel local officials to anticipate their preferences and solicit
their support.

Emerging Entrepreneurial Policy Climates

Revenue diversification trends shifting efforts away from grant- and
tax-supported activities to activities supported by fees, charges, and
innovative financing practices signal the emergence of a new entrepreneurial politics in cities (Marshall & Kirlin, 1988). In fact, there is
a remarkably similar emphasis on local economic development policies
in each of these countries, independent of the type of central-local
structure or the severity of fiscal stress. The Americans, French,
Danish, Norwegians, and Dutch all report that local officials are responding to declining national support by turning to local economic
development activities. While this is a common feature of American
local politics, it is more remarkable in the European context since a
common assumption has been that the more centralized European
public finance systems made such local activities and interjurisdiction-
al competition less necessary. The relative shift to more decentralized
structures — whether through formal decentralization policies or as a

consequence of national austerity measures — contributes to these convergent policy responses. This unexpected turn to local economic development has created a "vacuum" in central-local relationships that is only beginning to be charted (Young, 1986); the initial national response in the United States and Europe is benign but uncertain.

Qualitative Changes in Local Democracy

This new era of local fiscal management hints at qualitative changes in local democracy. On the one hand, there are few participation structures that give citizens access to fiscal policy decisions. Organized expenditure-demanding groups continue to be influential where they maintain alliances, as in France and Denmark, with local bureaucracies. In many instances, such linkages continue to shape and organize local political demands. Yet these very bureaucratic organizations are threatened by both ideological shifts toward privatization of public functions and basic changes in production systems that render them ill suited to decentralized, flexible, rapidly changing economies. The cyclical reemergence (Shefter, 1985) of taxpayers seeking political means to protect their disposable income and wealth in these uncertain economic times is hardly surprising, nor are the efforts to incorporate these amorphous interests in new local political coalitions. But political parties and elections, the traditional modes for articulating such concerns, appear to be imperfect means for transmitting citizens' preferences for responding to fiscal austerity. As the French experience illustrates, they allow for articulation of dissatisfaction but provide shaky grounds for issue formulation or political accountability (Kantor, 1988, p. 178). Coping with this imperfect fit between fiscal policy demands and local institutions for popular control is likely to be a lasting political consequence of local fiscal policy changes.

Notes

1. The fiscal dependence implied by total central government grants as a percentage of municipal revenues is a commonly used indirect indicator of local autonomy across systems, but not an uncontroversial one. Wolman (1982, p. 179) argues that the type of grant may influence the actual extent of central control and influence; limited data availability precludes breakdowns by grant type in this analysis. But note that Denmark and the United States are the only countries of the six studied that experienced declines in per capita total grants from 1978 to 1985 (in 1978 dollars). Of the four countries for which data are available, France and Norway report receiving more general grants per capita than conditional grants in 1984 (in 1978 dollars), while the United States reports

insignificant differences between the two types (.2) and Denmark reports more conditional grants than general grants per capita. In the United States, the cuts in general and conditional grants were similar, but in Denmark general grants were cut nearly in half while conditional grants actually increased. In Norway and France, both types of grants increased, but general grants increased at a higher rate.

2. The fiscal stress index used here reflects the revenues (grants and taxes) needed in the base year, taking into account inflation and changing population needs, to provide the level of services of two years previously without raising taxes (see Mouritzen & Nielsen, 1988; Wolman & Davis, 1980).

3. To compare fiscal management strategies, Mouritzen and Nielsen (1988) converted individual country frequencies to an "index of importance"; if all local officials in a country listed the strategy as "very important," the score would be 100 (p. 8). The lower the score, the less agreement on its importance. In using this index, the pattern of strategy rankings rather than the index scores themselves are the key features. In the tables in this chapter, index scores rather than country frequencies are reported to facilitate comparisons; these figures do not correspond, therefore, to the frequencies reported in the individual country chapters.

References

Ashford, D. (1979). Territorial politics and equality: Decentralization in the modern state. *Political Studies, 27*(1), 71-83.

Balme, R., Becquart-Leclercq, J., Clark, T. N., Hoffmann-Martinot, V., & Nevers, J. Y. (1987). New mayors: France and the U.S. *Tocqueville Review, 8*, 263-278.

Clark, G. L. (1984). A theory of local autonomy. *Annals of the Association of American Geographers, 74*, 195-208.

Clark, T. N., Balme, R., & Miranda, R. (1988). *Leadership patterns in American cities: Modeling policy processes.* Paper presented at the annual meeting of the American Political Science Association, Washington, DC.

Clark, T. N., & Ferguson, L. C. (1983). *City money: Political processes, fiscal strain, and retrenchment.* New York: Columbia University Press.

Elkin, S. (1985). Twentieth century urban regimes. *Journal of Urban Affairs, 7*(2), 11-28.

Fainstein, S., & Fainstein, N. (1983). Regime strategies, communal resistance, and economic forces. In S. Fainstein, N. Fainstein, et al. (Eds.), *Restructuring the city* (pp. 245-282). New York: Longman.

Gurr, T. R., & King, D. S. (1987). *The state and the city.* Chicago: University of Chicago Press.

Kantor, P. (1988). *The dependent city.* Glenview, IL: Scott Foresman/Little, Brown.

Marshall, D. R., & Kirlin, J. J. (1985). The distributive politics of the new federal system: Who wins? Who loses? In C. R. Warren (Ed.), *Urban policy in a changing federal system: Proceedings of a symposium.* Washington, DC: National Academy Press.

Levine, C. H., & Posner, P. L. (1981). The centralizing effects of austerity on the intergovernmental system. *Political Science Quarterly, 96*, 67-85.

Mouritzen, P. E., & Nielsen, K. H. (1988). *Handbook of comparative urban fiscal data.* Odense: Odense University, Danish Data Archives.

Nevers, J.-Y. (1988). *Grants allocation to French cities: The role of political processes.* Paper presented at the annual meeting of the American Political Science Association, Washington, DC.

Newton, K. (1980). *Balancing the books*. London: Sage.

Peterson, P. (1981). *City limits*. Chicago: University of Chicago Press.

Piven, F. F., & Friedland, R. (1984). Public choice and private power: A theory of fiscal crisis. In A. M. Kirby, P. Knox, & S. Pinch (Eds.), *Public service provision and urban development* (pp. 390-420). London: Croom Helm.

Przeworski, J. F. (1986). Changing intergovernmental relations and urban economic development. *Environment and Planning C: Government and Policy, 4*, 423-438.

Rose, R., & Page, E. (1982). Chronic instability in fiscal systems. In R. Rose & E. Page (Eds.), *Fiscal stress in cities* (pp. 198-245). London: Cambridge University Press.

Rubin, I., & Rubin, H. R. (1986). Structural theories and urban fiscal stress. In M. Gottdiener (Ed.) *Cities in stress: A new look at the urban crisis*. Beverly Hills, CA: Sage.

Schick, A. (1983). Incremental budgeting in a decremental age. *Policy Sciences, 16*(1), 1-25.

Schick, A. (1988). Microbudgetary adaptations to fiscal stress. *Public Administration Review, 48*(1), 523-533.

Sharp, E. B., & Elkin, D. R. (1987). The impact of fiscal limitation: A tale of seven cities. *Public Administration Review, 47*(5), 385-392.

Shefter, M. (1985). *Fiscal crisis/political crisis: The collapse and revival of New York*. New York: Basic Books.

Swanstrom, T. (1985). *The crisis of growth politics: Cleveland, Kucinich, and the challenge of urban populism*. Philadelphia: Temple University Press.

Walzer, N., & Jones, W. (in press). Fiscal austerity strategies and policy outcomes: A comparison of the U.S. and European countries. In P. E. Mouritzen (Ed.), *Defending city welfare*. London: Sage.

Wolman, H. (1982). Local autonomy and intergovernmental finance in Britain and the United States. In R. Rose & E. Page (Eds.), *Fiscal stress in cities* (pp. 168-195). London: Cambridge University Press.

Wolman, H. (in press). Fiscal stress and central-local relations: The critical role of government grants. In P. E. Mouritzen (Ed.), *Defending city welfare*. London: Sage.

Wolman, H., & Davis, B. (1980). *Local government strategies to cope with fiscal stress*. Washington, DC: Urban Institute.

Wong, K. (1985). *Changing the limits of the city: Toward a binary approach to urban policymaking*. Paper presented at the annual meeting of the American Political Science Association, New Orleans.

Young, K. (1986). Economic development in Britain: A vacuum in central-local relations. *Environment and Planning C: Government and Policy, 4*, 439-450.

Appendix:
Fiscal Austerity and Urban Innovation
Project Questionnaires

INFORMATION FROM COUNCIL MEMBERS

Please answer these questions about your city council.

PART I. FISCAL POLICIES

Q1 In the last three years, how important was the professional staff as compared to elected officials in affecting the overall spending level of your city government? (Circle one number)
1 PROFESSIONAL STAFF LARGELY SET LEVEL
2 PROFESSIONAL STAFF SUGGEST APPROXIMATE LEVEL
3 PROFESSIONAL STAFF AND ELECTED OFFICIALS INCLUDING MAYOR HAVE ABOUT EQUAL INPUT IN SETTING LEVEL
4 ELECTED OFFICIALS SET APPROXIMATE LEVEL
5 ELECTED OFFICIALS LARGELY SET LEVEL
6 FEW ANNUAL CHANGES, PAST PATTERNS USUALLY FOLLOWED INCREMENTALLY
7 DON'T KNOW/NOT APPLICABLE

Q2 How about in allocating funds among departments? (Circle one number)
1 PROFESSIONAL STAFF LARGELY SET ALLOCATIONS
2 PROFESSIONAL STAFF SET APPROXIMATE ALLOCATIONS
3 PROFESSIONAL STAFF AND ELECTED OFFICIALS HAVE ABOUT EQUAL INPUT IN SETTING ALLOCATIONS
4 ELECTED OFFICIALS SET APPROXIMATE ALLOCATIONS
5 ELECTED OFFICIALS LARGELY SET ALLOCATIONS
6 FEW ANNUAL CHANGES, PAST PATTERNS USUALLY FOLLOWED INCREMENTALLY
7 DON'T KNOW/NOT APPLICABLE

Q3 How about in developing new fiscal management strategies, such as imposing user charges for swimming or contracting out for services like garbage collection? (Circle one number)
1 PROFESSIONAL STAFF LARGELY DECIDE
2 PROFESSIONAL STAFF OFTEN DECIDE
3 PROFESSIONAL STAFF AND ELECTED OFFICIALS HAVE ABOUT EQUAL INPUT IN DEVELOPING NEW STRATEGIES
4 ELECTED OFFICIALS OFTEN DECIDE
5 ELECTED OFFICIALS LARGELY DECIDE
6 DON'T KNOW/NOT APPLICABLE

Some questions below refer to the "average" council member. Please do *not* include the mayor in considering these questions. Since councils often differ dramatically from each other, your best estimate of the "average" council member can still help place the council compared to others.

Q4 Please indicate the preference of the *average councilmember. Circle* one of the six answers for each of the 13 policy areas.

1 Spend *a lot less* on services provided by the city
2 Spend *somewhat less*
3 Spend *the same* as is now spent
4 Spend *somewhat more*
5 Spend *a lot more*
DK Don't know/not applicable

Q5 Please estimate the preference of the *majority of voters* in your city. Again, *circle* one of the six answers for each policy area.

1 Spend *a lot less* on services provided by the city
2 Spend *somewhat less*
3 Spend *the same* as is now spent
4 Spend *somewhat more*
5 Spend *a lot more*
DK Don't know/not applicable

Q6 Please indicate how often the *city government responded* favorably to the preference of the average council member. Circle one of the six answers for each policy area.

1 Almost never
2 Less than half the time
3 About half the time
4 More than half the time
5 Almost all the time
DK Don't know/not applicable

Policy Areas

		Q4	Q5	Q6
1.	ALL AREAS OF CITY GOVERNMENT	1 2 3 4 5 DK	1 2 3 4 5 DK	1 2 3 4 5 DK
2.	PRIMARY AND SECONDARY EDUCATION	1 2 3 4 5 DK	1 2 3 4 5 DK	1 2 3 4 5 DK
3.	SOCIAL WELFARE	1 2 3 4 5 DK	1 2 3 4 5 DK	1 2 3 4 5 DK
4.	STREETS AND PARKING	1 2 3 4 5 DK	1 2 3 4 5 DK	1 2 3 4 5 DK
5.	MASS TRANSPORTATION	1 2 3 4 5 DK	1 2 3 4 5 DK	1 2 3 4 5 DK

6. PUBLIC HEALTH AND HOSPITALS	1 2 3 4 5 DK	1 2 3 4 5 DK	1 2 3 4 5 DK
7. PARKS AND RECREATION	1 2 3 4 5 DK	1 2 3 4 5 DK	1 2 3 4 5 DK
8. LOW-INCOME HOUSING	1 2 3 4 5 DK	1 2 3 4 5 DK	1 2 3 4 5 DK
9. POLICE PROTECTION	1 2 3 4 5 DK	1 2 3 4 5 DK	1 2 3 4 5 DK
10. FIRE PROTECTION	1 2 3 4 5 DK	1 2 3 4 5 DK	1 2 3 4 5 DK
11. CAPITAL STOCK (e.g., ROADS, SEWERS, ETC.)	1 2 3 4 5 DK	1 2 3 4 5 DK	1 2 3 4 5 DK
12. NUMBER OF MUNICIPAL EMPLOYEES	1 2 3 4 5 DK	1 2 3 4 5 DK	1 2 3 4 5 DK
13. SALARIES OF MUNICIPAL EMPLOYEES	1 2 3 4 5 DK	1 2 3 4 5 DK	1 2 3 4 5 DK

PART II. PARTICIPANTS

Q7 Please indicate your judgment about the *spending preferences* of several participants in city government. *Circle* one of the six answers for each of the types of participants. Does the participant want to

1. Spend *a lot less* on services provided by the city
2. Spend *somewhat less*
3. Spend *the same* as is now spent
4. Spend *somewhat more*
5. Spend *a lot more*

DK Don't know/not applicable

Q8 Please indicate *how active* the participant has been in pursuing this spending preference. *Circle* one of the six answers for each of the types of participants. Has the participant carried on

1. No activity
2. Little activity
3. Some activity
4. A lot of activity
5. The most activity of any participant

DK Don't know/not applicable

Q9 Please indicate how often the *city government responded* favorably to the spending preferences of the participant in the last three years. *Circle* one of the six answers for each of the types of participants. The city has responded favorably

1. Almost never
2. Less than half the time
3. About half the time
4. More than half the time
5. Almost all the time

DK Don't know/not applicable

Participants

	Q7	Q8	Q9
1. PUBLIC EMPLOYEES AND THEIR UNIONS OR ASSOCIATIONS	1 2 3 4 5 DK	1 2 3 4 5 DK	1 2 3 4 5 DK
2. ORGANIZATIONS CONCERNED WITH LOW-INCOME GROUPS AND FAMILIES	1 2 3 4 5 DK	1 2 3 4 5 DK	1 2 3 4 5 DK
3. HOMEOWNERS' GROUPS OR ORGANIZATIONS	1 2 3 4 5 DK	1 2 3 4 5 DK	1 2 3 4 5 DK
4. NEIGHBORHOOD GROUPS OR ORGANIZATIONS	1 2 3 4 5 DK	1 2 3 4 5 DK	1 2 3 4 5 DK
5. CIVIC GROUPS (e.g., THE LEAGUE OF WOMEN VOTERS)	1 2 3 4 5 DK	1 2 3 4 5 DK	1 2 3 4 5 DK
6. ORGANIZATIONS CONCERNED WITH MINORITY GROUPS	1 2 3 4 5 DK	1 2 3 4 5 DK	1 2 3 4 5 DK

		Rating				

7. TAXPAYERS' ASSOCIATIONS   1 2 3 4 5 DK   1 2 3 4 5 DK   1 2 3 4 5 DK

8. BUSINESSMEN AND BUSINESS-ORIENTED GROUPS OR ORGANIZATIONS (e.g., CHAMBER OF COMMERCE)   1 2 3 4 5 DK   1 2 3 4 5 DK   1 2 3 4 5 DK

9. LOCAL MEDIA   1 2 3 4 5 DK   1 2 3 4 5 DK   1 2 3 4 5 DK

10. THE ELDERLY   1 2 3 4 5 DK   1 2 3 4 5 DK   1 2 3 4 5 DK

11. CHURCHES AND RELIGIOUS GROUPS   1 2 3 4 5 DK   1 2 3 4 5 DK   1 2 3 4 5 DK

12. INDIVIDUAL CITIZENS   1 2 3 4 5 DK   1 2 3 4 5 DK   1 2 3 4 5 DK

13. DEMOCRATIC PARTY   1 2 3 4 5 DK   1 2 3 4 5 DK   1 2 3 4 5 DK

14. REPUBLICAN PARTY   1 2 3 4 5 DK   1 2 3 4 5 DK   1 2 3 4 5 DK

15. MAYOR   1 2 3 4 5 DK   1 2 3 4 5 DK   1 2 3 4 5 DK

16. CITY COUNCIL   1 2 3 4 5 DK   1 2 3 4 5 DK   1 2 3 4 5 DK

17. CITY MANAGER or CAO   1 2 3 4 5 DK   1 2 3 4 5 DK   1 2 3 4 5 DK

18. CITY FINANCE STAFF   1 2 3 4 5 DK   1 2 3 4 5 DK   1 2 3 4 5 DK

19. DEPARTMENT HEADS   1 2 3 4 5 DK   1 2 3 4 5 DK   1 2 3 4 5 DK

20. FEDERAL AGENCIES   1 2 3 4 5 DK   1 2 3 4 5 DK   1 2 3 4 5 DK

21. STATE AGENCIES   1 2 3 4 5 DK   1 2 3 4 5 DK   1 2 3 4 5 DK

Q10 The types of participants listed above are frequently active in city government affairs. Could you list the five types of participants which are most influential in *decisions affecting city fiscal matters* in the past three years? (Feel free to mention participants not on the above list.)

1 _____
2 _____
3 _____
4 _____
5 _____

Q11 What are the five most influential participants in *city government in general* (not just in fiscal matters)?

_____ A. Same five as in Q10.
_____ B. If any are different from Q10, please list all five.

1 _____
2 _____
3 _____
4 _____
5 _____

PART III. VIEWS ON GOVERNMENT AND SOCIAL POLICIES

These are some widely used questions. We plan to use them to compare (1) cities with each other, and (2) city leaders with national samples of citizens.

Q12 If your city government were given an increase in General Revenue Sharing equal to 20 percent of total local expenditures, how would the average council member like to see the funds used? (Circle one number)
1 REDUCE PROPERTY TAXES AND OTHER LOCAL TAXES
2 INCREASE ALL SERVICES EQUALLY
3 INCREASE BASIC SERVICES (LIKE POLICE AND FIRE)
4 INCREASE CAPITAL CONSTRUCTION AND MAINTENANCE
5 INCREASE SOCIAL SERVICES
6 OTHER _____
_____
_____

Q13 "Federal income taxes should be cut by at least one third even if it means reducing military spending and cutting down on government services such as health and education." What does the average council member think? (Circle one number)
1 AGREE
2 DISAGREE
3 DON'T KNOW/NOT APPLICABLE

Q14 Should city council members vote mainly to do what is best for their district or ward or to do what is best for the city as a whole even if it doesn't really help their own district? What does the average council member think? (Circle one number)
1 DISTRICT
2 CITY AS WHOLE
3 DON'T KNOW

Q15 How important is the religious, ethnic, or national background of candidates in slating and campaigns for your city council? What does the average council member think? (Circle one number)

1 VERY IMPORTANT AND EXPLICITLY DISCUSSED IN SLATING AND CAMPAIGNS
2 FAIRLY IMPORTANT
3 OCCASIONALLY IMPORTANT
4 SELDOM IMPORTANT
5 VIRTUALLY NEVER EXPLICITLY DISCUSSED
6 DON'T KNOW/NOT APPLICABLE

Q16 Would the average council member favor or oppose a law which would require a person to obtain a police permit before he or she could buy a gun? (Circle one number)

1 FAVOR
2 OPPOSE
3 DON'T KNOW

Q17 In general, does the average council member favor or oppose the busing of black and white schoolchildren from one school district to another? (Circle one number)

1 FAVOR
2 OPPOSE
3 DON'T KNOW

Q18 "If a political leader helps people who need it, it doesn't matter that some of the rules are broken." What does the average council member think? (Circle one number)

1 AGREE
2 DISAGREE
3 DON'T KNOW

Q19 Would the average council member be for or against sex education in public schools? (Circle one number)

1 FOR
2 AGAINST
3 DON'T KNOW

Q20 Does the average council member think abortion should be legal under any circumstances, legal only under certain circumstances, or never legal under any circumstances? (Circle one number)

1 UNDER ANY CIRCUMSTANCES
2 UNDER CERTAIN CIRCUMSTANCES
3 NEVER LEGAL
4 DON'T KNOW/NOT APPLICABLE

Q21 During the coming months, President Reagan says he will try to hold down
government spending and taxes. Many congressmen, on the other hand, say
that Congress should pass social programs that would give more money to the
poor, the aged, and to schools and the like. Which position does the average
council member agree with more — holding down taxes and spending or
spending more money for social programs? (Circle one number)

1   HOLDING DOWN SPENDING
2   MORE SOCIAL PROGRAMS
3   UNDECIDED

PART IV. BACKGROUND INFORMATION

Finally, a few background questions for statistical purposes.

Q22 Although many cities have non-partisan elections, parties still are sometimes important. What political party, if any, do council members identify with? (How many of the council members are in each of the categories below.)

_____ 1 Republican
_____ 2 Democrat
_____ 3 None (Independent)
_____ 4 Don't know/Not applicable

Q23 How often did council members mention their party affiliations in their last campaign?

	A. The most partisan council member (Circle one number)	B. The least partisan council member (Circle one number)	C. The average partisan council member (Circle one number)
Almost always	1	1	1
Frequently	2	2	2
Seldom	3	3	3
Almost never	4	4	4
Never	5	5	5
Don't know/ Not applicable	6	6	6

Q24 How active are the two parties for average council elections? (Circle one number for each party)

1	2	3	4	5	6	7
DEMOCRATIC PARTY HELPS SELECT, ENDORSES, AND CAMPAIGNS ACTIVELY			DEMOCRATIC PARTY OCCASIONALLY ACTIVE IN CAMPAIGN			DEMOCRATIC PARTY NOT ACTIVE IN CAMPAIGN

1	2	3	4	5	6	7
REPUBLICAN PARTY HELPS SELECT, ENDORSES, AND CAMPAIGNS ACTIVELY			REPUBLICAN PARTY OCCASIONALLY ACTIVE IN CAMPAIGN			REPUBLICAN PARTY NOT ACTIVE IN CAMPAIGN

Q25 How many terms have the following council members served?

	A. The one who has served the most years (Circle one number)	B. The one who has served the least years (Circle one number)	C. The average council member (Circle one number)
One	1	1	1
Two	2	2	2
Three	3	3	3
More than three	4	4	4
Don't know/ Not applicable	5	5	5

Q26 How often did council members use the local media (radio, TV, the press) in the last two months of their last campaign? (Please include both paid advertisements and other media coverage.)

	A. The most visible council member (Circle one number)	B. The least visible council member (Circle one number)	C. The average council member (Circle one number)
Name appeared in the media several times each day	1	1	1
Name appeared about once/day	2	2	2
Name appeared a few times a week	3	3	3
Name appeared about once/week	4	4	4
Name appeared less than once/week		5	5
Don't know/ Not applicable	6	6	6

Q27 *Excluding* election periods, how often have council members appeared in the local media during the past two years?

	A. The most visible council member (Circle one number)	B. The least visible council member (Circle one number)	C. The average council member (Circle one number)
Name appeared in the media several times each day	1	1	1
Name appeared about once/day	2	2	2
Name appeared a few times a week	3	3	3
Name appeared about once/week	4	4	4
Name appeared less than once/week	5	5	5
Don't know/ Not applicable	6	6	6

Q28 Sometimes elected officials believe that they should take policy positions which are unpopular with the majority of their constituents. About how often would you estimate that council members vote *against* the dominant position of their constituents?

	A. The most citizen-responsive councilmember (Circle one number)	B. The least citizen-responsive councilmember (Circle one number)	C. The average councilmember (Circle one number)
Never or almost never	1	1	1
Only rarely	2	2	2
About once a month	3	3	3
More than once a month	4	4	4

Q29 How do council members feel about the total local tax burden (from city, county, school district, and special district governments)?

	A. The most anti-tax councilmember (Circle one number)	B. The most pro-tax councilmember (Circle one number)	C. The average councilmember (Circle one number)
Should be substantially reduced	1	1	1
Should be reduced somewhat	2	2	2
About right	3	3	3
Should be increased somewhat	4	4	4
Should be increased substantially	5	5	5
Don't know/ Not applicable	6	6	6

Q30 What is the age

	Of the oldest councilmember (Circle one number)	Of the youngest councilmember (Circle one number)	Of the average councilmember (Circle one number)
Under 20	1	1	1
21-30	2	2	2
31-40	3	3	3
41-50	4	4	4
51-60	5	5	5
Over 60	6	6	6

Q31 Approximately how many years of schooling did council members complete?

	A. The councilmember with the most years of schooling (Circle one number)	B. The councilmember with the fewest years of schooling (Circle one number)	C. The average councilmember (Circle one number)
None	1	1	1
Elementary school	2	2	2
High school	3	3	3
College	4	4	4
Graduate or professional school	5	5	5
Don't know/ Not applicable	6	6	6

Q32 Besides holding elective office, what are the major occupations of council
members? (How many of the council members are in each of the categories
below?)

_____ 1 No other occupation
_____ 2 Professional/technical
_____ 3 Managers and administrators
_____ 4 Sales and service
_____ 5 Clerical
_____ 6 Craftsmen
_____ 7 Laborers
_____ 8 Other
_____ 9 Don't know/not applicable

Q33 Are any council members current or former employees of the city government? (Circle one number)

1   NO
2   YES   If YES, how many council members _____ (Number)

Q34 Which of the following best describes the ethnic or racial identification of the
council members? (How many of the council members are in each of the categories below?)

_____ 1 Black
_____ 2 White (Caucasian)
_____ 3 Hispanic
_____ 4 Oriental
_____ 5 Other

Q35 What is the religious background of the council members? (How many of the
council members are in each of the categories below?)

_____ 1 Protestant
_____ 2 Catholic
_____ 3 Jewish
_____ 4 Don't know/Not applicable

Q36 How many of the council members are

_____ 1 Men
_____ 2 Women

Q37 Name of person completing this questionnaire. _____
    Address _____

_____

    Telephone: (___) _____

Is there anything else that you would like to tell us concerning Fiscal Austerity and Urban Innovation — any particular lessons from your city, solutions you may have found, or problems that remain that you would like to bring to the attention of others? Comments here or in a separate letter would be welcome.

Thanks for your help — we hope the results can help your city.

FISCAL AUSTERITY AND URBAN INNOVATION
October 18, 1982

CHIEF ADMINISTRATIVE OFFICER QUESTIONNAIRE

Q1   In the last three years, how important have each of the following problems been for your city's finances? (Circle one answer for each problem)

		One of Most Important	Very Important	Somewhat Important	Least Important	Don't Know/ Not applicable
1	Loss of Federal Revenue	MOST	VERY	SOMEWHAT	LEAST	DK
2	Loss of State Revenue	MOST	VERY	SOMEWHAT	LEAST	DK
3	Inflation	MOST	VERY	SOMEWHAT	LEAST	DK
4	Unemployment	MOST	VERY	SOMEWHAT	LEAST	DK
5	Declining Tax Base	MOST	VERY	SOMEWHAT	LEAST	DK
6	Rising Service Demands from Citizens	MOST	VERY	SOMEWHAT	LEAST	DK
7	State Tax, Revenue, or Expenditure Limits (like Proposition 13)	MOST	VERY	SOMEWHAT	LEAST	DK
8	Pressures from *Local* Taxpayers to Reduce Taxes and Spending	MOST	VERY	SOMEWHAT	LEAST	DK
9	Failure of Bond Referenda	MOST	VERY	SOMEWHAT	LEAST	DK
10	Mandated Costs from Federal and State Governments	MOST	VERY	SOMEWHAT	LEAST	DK
11	Pressures from Municipal Employees	MOST	VERY	SOMEWHAT	LEAST	DK
12	Other (Specify) _____					

Q2   What were your city's total expenditures (including operating and capital expenditures, paid fringe benefits and retirement contributions, and any enterprises like utilities operated by the municipal government) in

1980	1981	1982
_____	_____	_____

What were total revenues from all federal and state sources in

1980	1981	1982
_____	_____	_____

Q3 Does your state limit taxes, revenues, or expenditures of city governments? (Circle one number)

┌─ 1   NO
│ 2   YES ────────→ If YES, has your city reached this limit? (Circle one number)

1   YES, CITY HAS REACHED LIMIT
2   NO, BUT CLOSE TO LIMIT
3   NO, NOT CLOSE TO LIMIT

┌─────────────┐
│ GO TO Q7   │
└─────────────┘

→ Q4 Has this limit affected your fiscal practices? (Circle one number)

1   NO CLEAR EFFECT
2   SOME EFFECTS, BUT LIMITED
3   WE HAVE INCREASED BORROWING
4   WE HAVE CHANGED FEE SCHEDULES
5   WE HAVE FOUND NEW REVENUE SOURCES

→ Q5 Because of this limit, have any services or functions been particularly cut back? (Circle one number)

1   NO, NONE CUT BACK SIGNIFICANTLY
2   YES, SERVICES CUT ROUGHLY EQUALLY ACROSS ALL DEPARTMENTS
3   YES, SOME SERVICES WERE PARTICULARLY CUT. THESE WERE _____

_____

_____

(LIST SERVICES)

→ Q6 Were cutbacks made in these services which were (Circle as many as apply)

1   LEAST RESISTED BY MUNICIPAL EMPLOYEES
2   LEAST RESISTED BY SERVICE RECIPIENTS
3   LEAST RESISTED BY CITIZENS IN GENERAL
4   LEAST EFFICIENT
5   OTHER _____

_____

_____

_____

_____

Q7 Here is a list of fiscal management strategies that cities have used. Which have you used since January 1978? Please list the year when you *first implemented* each strategy that you have used. (Leave blank strategies that you have not used.)

Q8 Please indicate the importance in dollars of each strategy (since January 1978).

Strategy	Year implemented	One of most important	Very important	Somewhat important	Least important	Don't know/ not applicable
			(Circle your answer)			
Revenues						
1 Seek New Local Revenue Sources	19__	MOST	VERY	SOMEWHAT	LEAST	DK
2 Obtain Additional Intergovernmental Resources	19__	MOST	VERY	SOMEWHAT	LEAST	DK
3 Increase Taxes	19__	MOST	VERY	SOMEWHAT	LEAST	DK
4 Increase User Fees and Charges	19__	MOST	VERY	SOMEWHAT	LEAST	DK
5 Draw Down Surpluses	19__	MOST	VERY	SOMEWHAT	LEAST	DK
6 Sell Some Assets	19__	MOST	VERY	SOMEWHAT	LEAST	DK
7 Defer Some Payments to the Next Fiscal Year	19__	MOST	VERY	SOMEWHAT	LEAST	DK
8 Increase Short-Term Borrowing	19__	MOST	VERY	SOMEWHAT	LEAST	DK
9 Increase Long-Term Borrowing	19__	MOST	VERY	SOMEWHAT	LEAST	DK
Expenditures						
10 Impose Across-the-Board Cuts in All Departments	19__	MOST	VERY	SOMEWHAT	LEAST	DK
11 Cut Budget of Least Efficient Departments	19__	MOST	VERY	SOMEWHAT	LEAST	DK
12 Lay Off Personnel	19__	MOST	VERY	SOMEWHAT	LEAST	DK
13 Shift Responsibilities to Other Units of Government	19__	MOST	VERY	SOMEWHAT	LEAST	DK

14	Contract Out Services with Other Units of Government	19__	MOST	VERY	SOMEWHAT	LEAST	DK
15	Contract Out Services with Private Sector	19__	MOST	VERY	SOMEWHAT	LEAST	DK
16	Reduce Administrative (but not service) Expenditures	19__	MOST	VERY	SOMEWHAT	LEAST	DK
17	Reduce Employee Compensation Levels	19__	MOST	VERY	SOMEWHAT	LEAST	DK
18	Institute a Freeze on Wages and Salaries	19__	MOST	VERY	SOMEWHAT	LEAST	DK
19	Impose a Hiring Freeze	19__	MOST	VERY	SOMEWHAT	LEAST	DK
20	Reduce Work Force through Attrition	19__	MOST	VERY	SOMEWHAT	LEAST	DK
21	Reduce Expenditures for Supplies, Equipment, Travel	19__	MOST	VERY	SOMEWHAT	LEAST	DK
22	Reduce Services Funded by Own Revenues	19__	MOST	VERY	SOMEWHAT	LEAST	DK
23	Reduce Services Funded by Intergovernmental Revenues	19__	MOST	VERY	SOMEWHAT	LEAST	DK
24	Improve Productivity through Better Management	19__	MOST	VERY	SOMEWHAT	LEAST	DK
25	Improve Productivity by Adopting Labor-Saving Techniques	19__	MOST	VERY	SOMEWHAT	LEAST	DK
26	Eliminate Programs	19__	MOST	VERY	SOMEWHAT	LEAST	DK
27	Reduce Capital Expenditures	19__	MOST	VERY	SOMEWHAT	LEAST	DK
28	Keep Expenditure Increases Below Rate of Inflation	19__	MOST	VERY	SOMEWHAT	LEAST	DK
29	Early Retirements	19__	MOST	VERY	SOMEWHAT	LEAST	DK
30	Reduce Overtime	19__	MOST	VERY	SOMEWHAT	LEAST	DK
31	Joint Purchasing Agreements	19__	MOST	VERY	SOMEWHAT	LEAST	DK
32	Deferred Maintenance of Capital Stock	19__	MOST	VERY	SOMEWHAT	LEAST	DK
33	Impose Controls on New Construction to Help Limit Population Growth	19__	MOST	VERY	SOMEWHAT	LEAST	DK
34	Other (Specify) ___	19__	MOST	VERY	SOMEWHAT	LEAST	DK

Q9 Where would you place your city on the following items? (Circle the number most appropriate for your city)

	1	2	3	4	5
a. Revenue Forecasting	NO FORMAL FORECAST. WE USE LAST YEAR PLUS OR MINUS AN INCREMENT.		A SEPARATE FORECAST FOR EACH REVENUE SOURCE, WITH EXPLICIT CRITERIA FOR EACH SOURCE.		MULTI-YEAR AS WELL AS ANNUAL FORECASTING, USING COMPUTER SOFTWARE
b. Fiscal Information System	DEPARTMENT EXPENDITURES CENTRALLY MONITORED QUARTERLY OR LESS.		DEPARTMENT EXPENDITURES MONITORED AT LEAST MONTHLY AND DEPARTURES FROM BUDGET QUESTIONED BY FINANCE STAFF.		COMPUTERIZED SYSTEM IS USED TO MONITOR SPENDING BY ALL DEPARTMENTS ON A WEEKLY BASIS
c. Performance Measure	NO PERFORMANCE MEASURES USED.		FAIRLY SPECIFIC WORK LOAD MEASURE (E.G., TONS OF GARBAGE COLLECTED; HOURS OF POLICE PATROL) AND SOME EFFECTIVENESS MEASURES (E.G., PERCENT OF CITIZEN REQUESTS COVERED; POLICE RESPONSE TIME).		ALL DEPARTMENTS USE WORK LOAD AND EFFECTIVENESS MEASURES ANNUALLY AND COMPUTE COSTS OF SERVICE PROVISION ON A REGULAR BASIS
d. Accounting and Financial Reporting	MEET BUT DO NOT EXCEED THE REQUIREMENTS OF STATE AND LOCAL LAW.		CHANGE ACCOUNTING PROCEDURES ONLY AT SPECIFIC RECOMMENDATION OF OUTSIDE AUDITORS.		HELD MFOA CERTIFICATE OF CONFORMANCE FOR OVER FIVE OF LAST TEN YEARS.

	1	2	3	4	5
e. Management Rights	NON-MANAGEMENT EMPLOYEES, THROUGH THEIR REPRESENTATIVES OR CONTRACTS, HAVE FORMALIZED INFLUENCE ON MOST DECISIONS CONCERNING WORK SCHEDULES, CREATING OR ELIMINATING POSITIONS AND LAYOFFS.		MANAGEMENT MUST CONSULT WITH EMPLOYEE REPRESENTATIVES IN ABOUT HALF THESE DECISIONS.		MANAGEMENT POSSESSES THE SOLE RIGHT TO MAKE THESE DECISIONS.
f. Economic Development Impacts	NO SPECIFIC ANALYSIS DONE OF IMPACTS CHANGING ECONOMIC BASE.		OCCASIONAL STUDIES CONDUCTED OF MAJOR PROJECTS (E.G., PROJECTED COSTS AND BENEFITS TO CITY OF SHOPPING CENTER).		ECONOMIC BASE SYSTEMATICALLY MONITORED FOR CHANGES IN RETAIL SALES, JOBS CREATED OR LOST. SEVERAL ANALYSES OF SPECIFIC PROJECTS CONDUCTED EACH YEAR.
g. Reporting to Council	A REPORT IS GIVEN TO THE COUNCIL ON REVENUES AND BUDGET AT BUDGET TIME, JUST ONCE A YEAR		FINANCIAL REPORTS MADE SEVERAL TIMES DURING YEAR.		FINANCIAL REPORT MADE AT VIRTUALLY EVERY COUNCIL MEETING

Q10 Name of person completing this questionnaire. _____
       Address _____
       _____
       _____
       Telephone: (___)_____

Is there anything else that you would like to tell us concerning Fiscal
Austerity and Urban Innovation — any particular lessons from your city,
solutions you may have found, or problems that remain that you would like to
bring to the attention of others? Comments here or in a separate letter would
be welcomed.

Thanks for your help — we hope the results can help your city.

FISCAL AUSTERITY AND URBAN INNOVATION

October 18, 1982

INFORMATION FROM CITY MAYORS

PART I. FISCAL POLICIES

Q1 In the last three years how important was the professional staff as compared to elected officials in affecting the overall spending level of your city government? (Circle one number)

1 PROFESSIONAL STAFF LARGELY SET LEVEL
2 PROFESSIONAL STAFF SUGGEST APPROXIMATE LEVEL
3 PROFESSIONAL STAFF AND ELECTED OFFICIALS INCLUDING MAYOR HAVE ABOUT EQUAL INPUT IN SETTING LEVEL
4 ELECTED OFFICIALS SET APPROXIMATE LEVEL
5 ELECTED OFFICIALS LARGELY SET LEVEL
6 FEW ANNUAL CHANGES, PAST PATTERNS USUALLY FOLLOWED INCREMENTALLY
7 DON'T KNOW/NOT APPLICABLE

Q2 How about in allocating funds among departments? (Circle one number)

1 PROFESSIONAL STAFF LARGELY SET ALLOCATIONS
2 PROFESSIONAL STAFF SET APPROXIMATE ALLOCATIONS
3 PROFESSIONAL STAFF AND ELECTED OFFICIALS HAVE ABOUT EQUAL INPUT IN SETTING LEVEL
4 ELECTED OFFICIALS SET APPROXIMATE ALLOCATIONS
5 ELECTED OFFICIALS LARGELY SET ALLOCATIONS
6 FEW ANNUAL CHANGES, PAST PATTERNS USUALLY FOLLOWED INCREMENTALLY
7 DON'T KNOW/NOT APPLICABLE

Q3 How about in developing new fiscal management strategies, such as imposing user charges for swimming or contracting out for services like garbage collection? (Circle one number)

1 PROFESSIONAL STAFF LARGELY DECIDE
2 PROFESSIONAL STAFF OFTEN DECIDE
3 PROFESSIONAL STAFF AND ELECTED OFFICIALS HAVE ABOUT EQUAL INPUT IN DEVELOPING NEW STRATEGIES
4 ELECTED OFFICIALS OFTEN DECIDE
5 ELECTED OFFICIALS LARGELY DECIDE
6 DON'T KNOW/NOT APPLICABLE

Q4 Please indicate *your own preferences* about spending. *Circle* one of the six answers for each of the 13 policy areas.

1 Spend *a lot less* on services provided by the city
2 Spend *somewhat less*
3 Spend *the same* as is now spent
4 Spend *somewhat more*
5 Spend *a lot more*
DK Don't know/not applicable

Q5 Please estimate the preference of the *majority of voters* in your city. Again, *circle* one of the six answers for each policy area.

1 Spend *a lot less* on services provided by the city
2 Spend *somewhat less*
3 Spend *the same* as is now spent
4 Spend *somewhat more*
5 Spend *a lot more*
DK don't know/not applicable

Q6 Please indicate how successful you have been in implementing *your own spending preferences* in this term of office. *Circle* one of the six answers for each policy area.

1 Very successful
2 Somewhat successful
3 Somewhat unsuccessful
4 Very unsuccessful
5 The policy area is not in the jurisdiction of the city government
DK Don't know/not applicable

Policy Areas

		Q4	Q5	Q6
1.	ALL AREAS OF CITY GOVERNMENT	1 2 3 4 5 DK	1 2 3 4 5 DK	1 2 3 4 5 DK
2.	PRIMARY AND SECONDARY EDUCATION	1 2 3 4 5 DK	1 2 3 4 5 DK	1 2 3 4 5 DK
3.	SOCIAL WELFARE	1 2 3 4 5 DK	1 2 3 4 5 DK	1 2 3 4 5 DK
4.	STREETS AND PARKING	1 2 3 4 5 DK	1 2 3 4 5 DK	1 2 3 4 5 DK
5.	MASS TRANSPORTATION	1 2 3 4 5 DK	1 2 3 4 5 DK	1 2 3 4 5 DK
6.	PUBLIC HEALTH AND HOSPITALS	1 2 3 4 5 DK	1 2 3 4 5 DK	1 2 3 4 5 DK
7.	PARKS AND RECREATION	1 2 3 4 5 DK	1 2 3 4 5 DK	1 2 3 4 5 DK

8.	LOW-INCOME HOUSING	1	2	3	4	5	DK	1	2	3	4	5	DK	1	2	3	4	5	DK
9.	POLICE PROTECTION	1	2	3	4	5	DK	1	2	3	4	5	DK	1	2	3	4	5	DK
10.	FIRE PROTECTION	1	2	3	4	5	DK	1	2	3	4	5	DK	1	2	3	4	5	DK
11.	CAPITAL STOCK (e.g., ROADS, SEWERS, ETC.)	1	2	3	4	5	DK	1	2	3	4	5	DK	1	2	3	4	5	DK
12.	NUMBER OF MUNICIPAL EMPLOYEES	1	2	3	4	5	DK	1	2	3	4	5	DK	1	2	3	4	5	DK
13.	SALARIES OF MUNICIPAL EMPLOYEES	1	2	3	4	5	DK	1	2	3	4	5	DK	1	2	3	4	5	DK

PART II. PARTICIPANTS

Q7 Please indicate your judgment about the *spending preferences* of several participants in city government affairs. *Circle* one of the six answers for each of the types of participants. Does the participant want to

1. Spend *a lot less* on services provided by the city
2. Spend *somewhat less*
3. Spend *the same* as is now spent
4. Spend *somewhat more*
5. Spend *a lot more*
   DK Don't know/not applicable

Q8 Please indicate *how active* the participant has been in pursuing this spending preference. *Circle* one of the six answers for each of the types of participants. Has the participant carried on

1. No activity
2. Little activity
3. Some activity
4. A lot of activity
5. The most activity of any participant
   DK Don't know/not applicable

Q9 Please indicate how often the *city government responded* favorably to the spending preference of the participant in the last three years. *Circle* one of the six answers for each of the types of participants. The city has responded favorably

1. Almost never
2. Less than half the time
3. About half the time
4. More than half the time
5. Almost all the time
   DK Don't know/not applicable

Participants

	Q7	Q8	Q9
1. PUBLIC EMPLOYEES AND THEIR UNIONS OR ASSOCIATIONS	1 2 3 4 5 DK	1 2 3 4 5 DK	1 2 3 4 5 DK
2. ORGANIZATIONS CONCERNED WITH LOW-INCOME GROUPS AND FAMILIES	1 2 3 4 5 DK	1 2 3 4 5 DK	1 2 3 4 5 DK
3. HOMEOWNERS' GROUPS OR ORGANIZATIONS	1 2 3 4 5 DK	1 2 3 4 5 DK	1 2 3 4 5 DK
4. NEIGHBORHOOD GROUPS OR ORGANIZATIONS	1 2 3 4 5 DK	1 2 3 4 5 DK	1 2 3 4 5 DK
5. CIVIC GROUPS (e.g., THE LEAGUE OF WOMEN VOTERS)	1 2 3 4 5 DK	1 2 3 4 5 DK	1 2 3 4 5 DK
6. ORGANIZATIONS CONCERNED WITH MINORITY GROUPS	1 2 3 4 5 DK	1 2 3 4 5 DK	1 2 3 4 5 DK

#	Item	Rating	Rating	Rating
7.	TAXPAYERS' ASSOCIATIONS	1 2 3 4 5 DK	1 2 3 4 5 DK	1 2 3 4 5 DK
8.	BUSINESSMEN AND BUSINESS-ORIENTED GROUPS OR ORGANIZATIONS (e.g., CHAMBER OF COMMERCE)	1 2 3 4 5 DK	1 2 3 4 5 DK	1 2 3 4 5 DK
9.	LOCAL MEDIA	1 2 3 4 5 DK	1 2 3 4 5 DK	1 2 3 4 5 DK
10.	THE ELDERLY	1 2 3 4 5 DK	1 2 3 4 5 DK	1 2 3 4 5 DK
11.	CHURCHES AND RELIGIOUS GROUPS	1 2 3 4 5 DK	1 2 3 4 5 DK	1 2 3 4 5 DK
12.	INDIVIDUAL CITIZENS	1 2 3 4 5 DK	1 2 3 4 5 DK	1 2 3 4 5 DK
13.	DEMOCRATIC PARTY	1 2 3 4 5 DK	1 2 3 4 5 DK	1 2 3 4 5 DK
14.	REPUBLICAN PARTY	1 2 3 4 5 DK	1 2 3 4 5 DK	1 2 3 4 5 DK
15.	MAYOR	1 2 3 4 5 DK	1 2 3 4 5 DK	1 2 3 4 5 DK
16.	CITY COUNCIL	1 2 3 4 5 DK	1 2 3 4 5 DK	1 2 3 4 5 DK
17.	CITY MANAGER or CAO	1 2 3 4 5 DK	1 2 3 4 5 DK	1 2 3 4 5 DK
18.	CITY FINANCE STAFF	1 2 3 4 5 DK	1 2 3 4 5 DK	1 2 3 4 5 DK
19.	DEPARTMENT HEADS	1 2 3 4 5 DK	1 2 3 4 5 DK	1 2 3 4 5 DK
20.	FEDERAL AGENCIES	1 2 3 4 5 DK	1 2 3 4 5 DK	1 2 3 4 5 DK
21.	STATE AGENCIES	1 2 3 4 5 DK	1 2 3 4 5 DK	1 2 3 4 5 DK

Q10 The types of participants listed above are frequently active in city government affairs. Could you list the five types of participants which are most influential in *decisions affecting city fiscal matters* in the past three years? (Feel free to mention participants not on the above list.)

1 _____
2 _____
3 _____
4 _____
5 _____

Q11 What are the five most influential participants in *city government in general* (not just in fiscal matters)?
_____ A. Same five as in Q10.
_____ B. If any are different from Q10, please list all five.

1 _____
2 _____
3 _____
4 _____
5 _____

PART III. VIEWS ON GOVERNMENT AND SOCIAL POLICIES

These are some widely used survey questions. We plan to use them to compare (1) cities with each other, and (2) city leaders with national samples of citizens.

Q12 If your city government were given an increase in General Revenue Sharing equal to 20 percent of total local expenditures, how would you like to see the funds used? (Circle one number)

1   REDUCE PROPERTY TAXES AND OTHER LOCAL TAXES
2   INCREASE ALL SERVICES EQUALLY
3   INCREASE BASIC SERVICES (LIKE POLICE AND FIRE)
4   INCREASE CAPITAL CONSTRUCTION AND MAINTENANCE
5   INCREASE SOCIAL SERVICES
6   OTHER _____
    _____
    _____
    _____

Q13 "Federal income taxes should be cut by at least one third even if it means reducing military spending and cutting down on government services such as health and education." What is your feeling? (Circle one number)

1   AGREE
2   DISAGREE
3   DON'T KNOW/NOT APPLICABLE

Q14 Should city council members vote mainly to do what is best for their district or ward or to do what is best for the city as a whole even if it doesn't really help their own district? What is your feeling? (Circle one number)

1   DISTRICT
2   CITY AS WHOLE
3   DON'T KNOW

Q15 How important is the religious, ethnic, or national background of candidates in slating and campaigns for your city council? (Circle one number)
1 VERY IMPORTANT AND EXPLICITLY DISCUSSED IN SLATING AND CAMPAIGNS
2 FAIRLY IMPORTANT
3 OCCASIONALLY IMPORTANT
4 SELDOM IMPORTANT
5 VIRTUALLY NEVER EXPLICITLY DISCUSSED
6 DON'T KNOW/NOT APPLICABLE

Q16 Would you favor or oppose a law which would require a person to obtain a police permit before he or she could buy a gun? (Circle one number)
1 FAVOR
2 OPPOSE
3 DON'T KNOW

Q17 In general, do you favor or oppose the busing of black and white school-children from one school to another? (Circle one number)
1 FAVOR
2 OPPOSE
3 DON'T KNOW/NOT APPLICABLE

Q18 "If a political leader helps people who need it, it doesn't matter that some of the rules are broken." What is your feeling? (Circle one number)
1 AGREE
2 DISAGREE
3 DON'T KNOW/NOT APPLICABLE

Q19 Would you be for or against sex education in public schools? (Circle one number)
1 FOR
2 AGAINST
3 DON'T KNOW/NOT APPLICABLE

Q20 Do you think abortion should be legal under any circumstances, legal only under certain circumstances, or never legal under any circumstances? (Circle one number)
1 UNDER ANY CIRCUMSTANCES
2 UNDER CERTAIN CIRCUMSTANCES
3 NEVER LEGAL
4 DON'T KNOW/NOT APPLICABLE

Q21 During the coming months, President Reagan says he will try to hold down government spending and taxes. Many congressmen, on the other hand, say that Congress should pass social programs that would give more money to the poor, the aged, and to schools and the like. Which position do you agree with more — holding down spending and taxes or spending more money for social programs? (Circle one number)
1 HOLDING DOWN SPENDING
2 MORE SOCIAL PROGRAMS
3 UNDECIDED

PART IV. BACKGROUND INFORMATION

Finally, a few background questions about yourself for statistical purposes.

Q22 How many terms have you served as mayor? (Circle one number)
1   One (this is my first term)
2   Two
3   Three
4   More than three

Q23 Approximately how many years have you spent in *elected* office?
_____ (years)

Q24 Although many cities have non-partisan elections, parties still are sometimes important. What political party, if any, do you identify with? (Circle one number)
1   Republican
2   Democrat
3   None (Independent)
4   Other (please specify)_____

Q25 How often did you mention this party affiliation in your last campaign? (Circle one number)
1   Almost always
2   Frequently
3   Seldom
4   Almost never
5   Never

Q26 How active was your party in your last election? (Circle one number)
1   Party helped select and endorse me and was active in campaign
2   Party active in campaign
3   Party occasionally participated
4   Party not active in campaign

Q27 Approximately how often do you meet with local party officials? (Circle one number)
1   Several times a month
2   Once a month
3   Several times a year
4   Seldom
5   Never

Q28 How often did you use the local media (radio, TV, the press) in the last two months of your last campaign? (Please include both paid advertisements and other media coverage.) (Circle one number)
1   Name appeared in the media several times a day
2   Name appeared about once a day
3   Name appeared about once a week
4   Name appeared less than once a week
5   Don't know/not applicable

Q29 *Excluding* election periods, how often have you appeared in the local media in the past two years? (Circle one number)

1 Name appeared in the media several times a day
2 Name appeared about once a day
3 Name appeared a few times a week
4 Name appeared about once a week
5 Name appeared less than once a week
6 Don't know/not applicable

Q30 Sometimes elected officials believe that they should take policy positions which are unpopular with the majority of their constituents. About how often would you estimate that you took a position *against* the dominant opinion of your constituents? (Circle one number)

1 Never or almost never
2 Only rarely
3 About once a month
4 More than once a month
5 Regularly

Q31 How do you feel about the total local tax burden (from city, county, school district, and special district governments)? (Circle one number)

1 Should be substantially reduced
2 Should be reduced somewhat
3 About right
4 Should be increased somewhat
5 Should be increased substantially
6 Don't know

Q32 Please indicate your present age.

_____ (years)

Q33 Which of the following best describes your ethnic or racial identification? (Circle one number)

1 Black
2 White (Caucasian)
3 Hispanic
4 Oriental
5 Other _____

Q34 Are you (Circle one number)

1 Male
2 Female

Q35 Is your religious background (Circle one number)

1 Protestant
2 Catholic
3 Jewish
4 Other

Q36 What is the highest grade or year in elementary school, high school, or college you have completed? (Circle one number)

None 1
Elementary 01 02 03 04 05 06 07 08
High school . . 09 10 11 12
College 13 14 15 16
Some graduate school . . . 17
Graduate or professional degree 18

Q10 Name of person completing this questionnaire. _____
    Address _____
    _____
    _____
    Telephone: (___)_____

Is there anything else that you would like to tell us concerning Fiscal Austerity and Urban Innovation — any particular lessons from your city, solutions you may have found, or problems that remain that you would like to bring to the attention of others? Comments here or in a separate letter would be welcome.

Thank you for your help — we hope the results can help your city.

Index